Kinetic Anatomy

Second Edition

Robert S. Behnke, HSD

Professor Emeritus, Indiana State University

Human Kinetics

Library of Congress Cataloging-in-Publication Data

Behnke, Robert S., 1938-
  Kinetic anatomy/ Robert S. Behnke.--2nd ed.
     p. ; cm.
  Includes bibliographical references and index.
  ISBN 0-7360-5909-1 (softcover)
  1. Musculoskeletal system--Anatomy. 2. Human locomotion. 3. Human anatomy.
  [DNLM: 1. Musculoskeletal Physiology. 2. Musculoskeletal System--anatomy & histology. 3.
Movement. WE 103 B419k 2006] I. Title.
   QM100.B445 2006
   611'.7--dc22

                              2005006622

ISBN-10: 0-7360-5909-1
ISBN-13: 978-0-7360-5909-1

Copyright © 2006, 2001 by Human Kinetics, Inc.

Acquisitions Editor: Loarn D. Robertson, PhD; **Developmental Editor:** Maggie Schwarzentraub; **Assistant Editors:** Amanda M. Eastin, Lee Alexander, and Maureen Eckstein; **Copyeditor:** Julie Anderson; **Proofreader:** Red Inc.; **Indexer:** Marie Rizzo; **Graphic Designer:** Bob Reuther; **Graphic Artist:** Denise Lowry; **Photo Manager:** Sarah Ritz; **Cover Designer:** Jack W. Davis; **Photographer (cover):** Jamie Squire/Getty Images; **Photographer (interior):** Tom Roberts, except where otherwise noted. Photos on pages 1 and 33 © Human Kinetics. Photo on page 117 © Photodisc. Photo on page 171 © Barry Giles. **Art Manager:** Kelly Hendren; **Medical Illustrator:** Jason M. McAlexander, MFA; **Printer:** Custom Color

Printed in the United States of America   10   9   8   7   6   5   4   3

Human Kinetics
Web site: www.HumanKinetics.com

United States: Human Kinetics, P.O. Box 5076, Champaign, IL 61825-5076
800-747-4457
e-mail: humank@hkusa.com

Canada: Human Kinetics, 475 Devonshire Road, Unit 100, Windsor, ON N8Y 2L5
800-465-7301 (in Canada only)
e-mail: orders@hkcanada.com

Europe: Human Kinetics, 107 Bradford Road, Stanningley
Leeds LS28 6AT, United Kingdom
+44 (0) 113 255 5665
e-mail: hk@hkeurope.com

Australia: Human Kinetics, 57A Price Avenue, Lower Mitcham, South Australia 5062
08 8372 0999
e-mail: liaw@hkaustralia.com

New Zealand: Human Kinetics, Division of Sports Distributors NZ Ltd.
P.O. Box 300 226 Albany, North Shore City, Auckland
0064 9 448 1207
e-mail: info@humankinetics.co.nz

At one time or another in our lives, we come in contact with someone whom we consider our teacher, supervisor, mentor, or our role model. In many instances, these people may also become our friends. I was fortunate during my professional preparation to have all of the above and more in one individual. During my professional preparation, this person emphasized the importance of knowing human anatomy, because he believed it was a keystone to understanding athletic performance and to preventing, recognizing, treating, and rehabilitating athletic trauma. This emphasis has inspired me throughout the preparation of this book. I'd like to thank Robert Nicolette, former head athletic trainer (1957–1969) at the University of Illinois, on behalf of all of us who are fortunate enough to know him. He has touched everyone we work with as a result of our association with him. I dedicate this book to him to express how much I appreciate him.

Contents

Preface

Human anatomy has not changed during our lifetimes, so why would a textbook titled *Kinetic Anatomy* need a second edition? Although the subject matter is essentially unchanged, the manner in which it is presented is always subject to change. Numerous faculty and students looking for an entry-level text positively received the first edition of *Kinetic Anatomy*. As a result of their comments and suggestions, a second edition of *Kinetic Anatomy* has been prepared in a manner that will further enhance learning. I thank all those who took the time to provide helpful feedback and suggestions.

Students studying human anatomy will find this second edition even more helpful, with new full-color illustrations that enhance the written text. The "hands-on" experiences that have readers use their own bodies or that of a partner to learn various anatomical structures and the "focus points" that present common anatomical conditions were well received in the first edition, and additional experiences have been included in the second edition. Also included from the first edition are the end-of-chapter reviews of terminology (lists of the key terms that are boldfaced within each chapter), suggested learning activities, and practice multiple-choice and fill-in-the-blank questions pertaining to material introduced in the chapter. An answer sheet for these questions appears after the final chapter of the text. Also retained from the first edition is a list of suggested readings for those who want additional information about material presented in this book. New to the second edition is an index that allows for easier use of the book.

The primary goals of *Kinetic Anatomy, Second Edition,* remain as they were in the first edition. One goal is to present the basic vocabulary of anatomy. This knowledge will enable readers to communicate with colleagues, physicians, therapists, educators, coaches, allied health personnel, and others using a universal language of human anatomy and enhanced comprehension of human anatomy. A second goal is to give readers a firm concept of how a human body is constructed and how it moves by discussing bones, tying the bones together to make articulations (joints), placing muscles on the bones (crossing joints), and then observing how the joints move when the muscles contract. The book also discusses the nerves and blood vessels that supply the muscles essential to movement, but the main emphasis is on putting together the human body for the purpose of studying movement. Knowing what structures are involved and how they should function allows an individual to identify problems and correct them to enhance physical activity. Other areas of human anatomy require more extensive learning of the internal organs, the nervous system, the cardiovascular system, the digestive system, and the respiratory system. The typical student of physical education, athletic training, and allied health care fields will likely examine these systems in advanced studies of anatomy, physiology, exercise physiology, biomechanics, neurophysiology, and other scientific offerings involving the human body. A third goal, possibly less academic but most important of all, is to impart knowledge that allows the pursuit of healthy living. Knowing about your body can alert you to potential problems and, with other acquired information, help you prevent or resolve those problems and lead a healthful lifestyle.

As it was in the first edition, this text is organized into four parts. Part I discusses the basic concepts of anatomy. The remainder of the text, like many textbooks on kinesiology and biomechanics, divides the body into the upper extremity (part II of this text); spinal column, pelvis, and thorax (part III); and lower extremity (part IV).

Each anatomical chapter in parts II, III, and IV follows the same format: bones, joints and ligaments, and muscles. Each of these parts concludes with a chapter examining the major nerves and blood vessels of the anatomical structures discussed in that part. The part II, III, and IV summary tables for muscles, nerves, and blood vessels that were so well received in the first edition have been supplemented in the second edition with summary tables of bones, joints, ligaments, movements.

Faculty choosing to use the second edition of *Kinetic Anatomy* have access to two ancillary products: an instructor guide, with chapter summaries, lecture aids, activities, and selected figures from the text that can be used to create a Microsoft PowerPoint® presentation; and a test package that includes questions (multiple-choice, true-and-false, fill-ins) for each chapter.

Students and instructors alike will benefit from the software program *Essentials of Interactive Functional Anatomy* by Primal Pictures that is bound into the back of this book. This learning tool allows the user to remove structures layer by layer through 11 layers (from skin to bone) with the "strip away" technique. Accompanying text includes information regarding proximal and distal attachments, nerve innervation, blood supply, and primary and secondary actions. Views of any specific structure on the screen can be rotated up to 360° and stopped at any point for viewing. Additionally, the live-action video option allows users to observe muscle actions during walking, standing from a sitting position, and other activities such as push-ups and sit-ups. A zoom control allows the user to zoom in for a closer look at specific details or zoom out for overall views of any structure. Every structure has related text material to further define the structure being viewed.

The anatomy text of *Essentials of Interactive Functional Anatomy* includes bones, ligaments, muscles, tendons, retinacula, capsules, cartilage, discs, membranes, and other miscellaneous structures. For *Kinetic Anatomy* users, the program includes detailed animations for the shoulder, elbow and forearm, wrist and hand, trunk, hip, knee, and ankle and foot. Each joint has a muscle-action analysis of each fundamental movement of the joint. *Essentials of Interactive Functional Anatomy* provides excellent learning opportunities for students of human anatomy, kinesiology, and biomechanics.

Both students and instructors will benefit from the additions to *Kinetic Anatomy* in this second edition. As was said in the first edition of *Kinetic Anatomy,* the study of human anatomy is a fascinating subject particularly because it is about *you.* Enjoy!

Acknowledgments

The second edition has been enhanced with new illustrations in each chapter, wonderfully produced by Jason M. McAlexander. This artist and subsequent artwork developed for this edition were strongly influenced as a result of a very unique experience. During the 1940s, 1950s, and early 1960s, all physical education majors at the University of Illinois were required to take 10 semester hours of Physiological Anatomy (two 5 semester hour courses). A wonderful professor of anatomy and physiology at the university designed and taught these two courses. Dr. Walter Phillipp Elhardt designed his own textbook for the course: *Physiological Anatomy: A Text-Book and Laboratory Manual for Physical Education Students.* The ninth (1959) and final edition was published by John S. Swift Company, Chicago, Illinois. The book was actually Dr. Elhardt's typed lectures combined with anatomical drawings primarily done by Thomas S. Stricker, a friend of Dr. Elhardt's and former B-25 pilot who was shot down over Germany during World War II. The illustrations new to *Kinetic Anatomy,* second edition, are a tribute to and an appreciation for these wonderful men. Their influence on hundreds of physical educators and coaches during their educational preparation will never be forgotten. I thank these men for their influence on and, above all, their wonderful inspiration for this edition of *Kinetic Anatomy.*

Several individuals at Human Kinetics Publishers (HK) have contributed tremendously to this edition of *Kinetic Anatomy.* First, Dr. Loarn Robertson, senior acquisitions editor, must be thanked for initially inspiring the development of *Kinetic Anatomy* and then insisting a second edition could be produced that would enhance the material presented in the first edition. Second, Maggie Schwarzentraub, developmental editor, earns my respect and appreciation for her guidance, suggestions, and persistence in seeing this project to its conclusion. Last, I thank Dr. Rainer Martens, HK president and founder, for approving the vast artwork project making the second edition of *Kinetic Anatomy* so beautifully illustrated. These three people not only made working on the second edition of *Kinetic Anatomy* a pleasure but also have produced a publication that will make the study of human anatomy enjoyable for anyone interested in learning about the human body and how it moves.

PART I

General Concepts of Anatomy

1

Structures

CHAPTER OUTLINE

Human anatomy has been defined simply as the structure of organisms pertaining to mankind. A structure is, by one definition, something composed of interrelated parts forming an organism, and an organism is simply defined as a living thing. The body is made up of four different types of tissues (a collection of a similar type of cells). **Connective tissue** makes up bone, cartilage, and soft tissue such as skin, fascia, tendons, and ligaments. **Muscle tissue** is divided into three types: skeletal, which moves the parts of the skeleton; cardiac, which causes the pumping action of the heart; and smooth, which lines arterial walls and other organs of the body. **Nerve tissue** is divided into neurons, which conduct impulses involving the brain, the spinal cord, spinal nerves, and cranial nerves, and **neuroglia,** which are specifically involved in the cellular processes that support the neurons both metabolically and physically. The fourth type of tissue is known as **epithelial tissue.** There are four varieties, and all are involved with the structures of the respiratory, gastrointestinal, urinary, and reproductive systems.

The study of human anatomy as it pertains to movement concentrates on the bones, joints (ligaments), and muscles responsible for the human body's movement. Additionally, the roles of the nervous system in stimulating muscle tissue and of the vascular system in providing the muscle tissue with energy and removing by-products need to be studied. Although anatomy also includes subjects such as the endocrine system, respiratory system, digestive system, reproductive system, autonomic nervous system, circulatory system, urinary system, and sensory organs, this text concentrates specifically on those anatomical structures chiefly responsible for making movement of the human organism possible.

Proper vocabulary is extremely important when discussing anatomy. Common terms make communication with others (physicians, coaches, therapists, athletic trainers) much easier, and it is essential that a student of human anatomy become familiar with standard terminology presented in this chapter. Knowledge of the structures and common terms used to describe movement anatomically also facilitates the use of specific coaching principles; the use of therapeutic techniques involving human movement for prevention, treatment, and rehabilitation of various physical conditions; and the application of scientific principles to human movement.

Although all systems of the human organism can be said to contribute in some unique way to movement, this text emphasizes those systems (skeletal, articular, muscular, nervous, and circulatory) that directly accomplish movement. Primary concentration is on the following structures: bones, ligaments, joints, and muscles producing movement, with additional comments about the nerves and blood vessels in each specific anatomical area.

Bones

The body contains 206 bones. Bones have several functions, such as support, protection, movement, mineral storage, and blood cell formation. Arrangements of bones that form joints and the muscular attachments to those bones determine movement. Bones are classified by their shapes into four groups: long bones, short bones, flat bones, and irregular bones. Some authors also distinguish a fifth type of bone, known as sesamoid bones, which are small, nodular bones embedded in tendon (figure 1.1). The bones that provide the framework for the body and that make movement possible are classified as **long bones** (figure 1.2). A long bone has a shaft, known as the **diaphysis,** and two large prominences at either end of the diaphysis, known as **epiphyses.** Early in life the epiphysis is separated from the diaphysis by a cartilaginous structure known as the **epiphyseal plate.** It is from these epiphyseal plates at both ends of the diaphysis that the bone grows; thus, this area is often referred to as the *growth plate.* Once a bone has reached its maximum length (maturity), the epiphyseal plate "closes" (bone tissue has totally replaced the cartilaginous tissue), and the epiphysis and diaphysis become one continuous structure. Around the entire bone is a layer of tissue known as the **periosteum,** where bone cells are produced. Additionally, the very ends of each bone's epiphyses are covered with a mate-

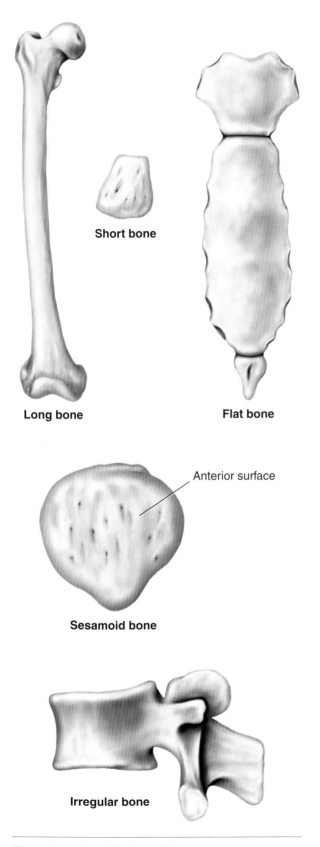

Figure 1.1   Classification of bones.

rial known as **articular cartilage.** This covering provides for smooth movement between the bones that make up a joint and protects the end of the bone from wear and tear.

Short bones differ from long bones in that they possess no diaphysis and are fairly symmetrical. Bones in the wrist and ankle are examples of short bones. Flat bones, such as the bones of the head, chest, and shoulder, get their name from their flat shape. Irregular bones are simply bones that cannot be classified as long, short, or flat. The best example of an irregular bone is a vertebra of the spinal column. An additional classification that some anatomists recognize is sesamoid (sesame seed–shaped) bones. These oval bones are free-floating bones usually found within tendons of muscles. The kneecap (patella) is the largest sesamoid bone in the body; others are found in the hand and the foot.

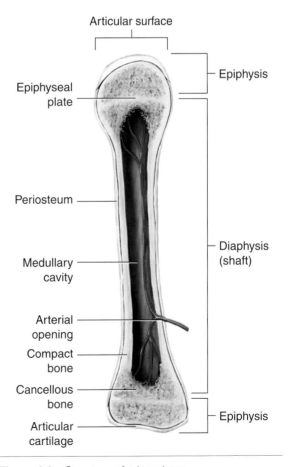

Figure 1.2   Structure of a long bone.

Several terms are commonly used to describe features of bones. These features are usually referred to as **anatomical landmarks** and are basic to one's anatomical vocabulary. A **tuberosity** on a bone is a large bump (figures 1.3 and 1.7). A **process** is a projection from a bone (figure 1.3). A **tubercle** is a smaller bump (figure 1.4). All three of these bony prominences usually serve as the attachment for other structures. A **spine,** or **spinous process,** is typically a longer and thinner

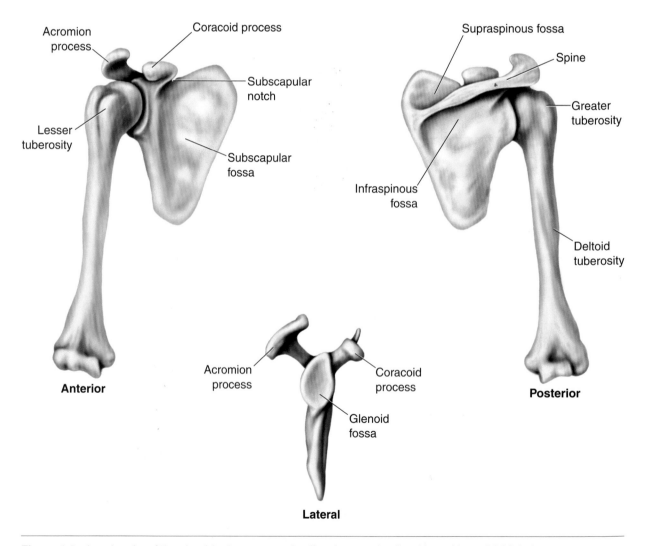

Figure 1.3   Landmarks of the shoulder bones: anterior (front), posterior (back), and lateral (side) views.

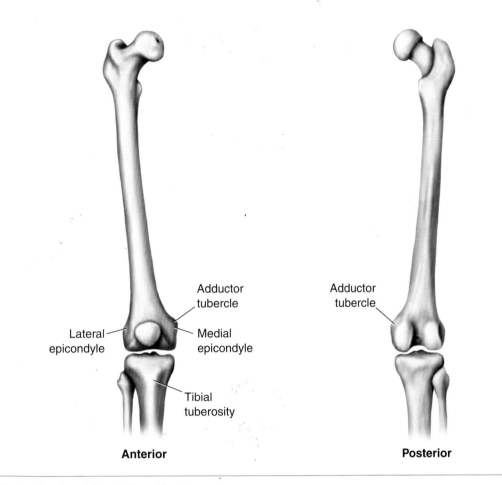

Figure 1.4 Landmarks of the thigh and leg bones.

projection of bone, unlike any of the previously mentioned prominences (figure 1.5). The large bony knobs at either end of a long bone are known as the **condyles** (figure 1.6). The part of the condyle that articulates (joins) with another bone is known as the **articular surface** (figure 1.2). Smaller bony knobs that sometimes appear just above the condyles of a bone are known as **epicondyles** (figure 1.4). A **fossa** is a smooth, hollow surface on a bone and usually functions as a source of attachment for other structures (figure 1.3). A smaller and flatter smooth surface is a **facet** (figure 1.8). Facets also serve for attachment of other structures. A **notch** is an area on a bone that appears to be cut out and allows for the passage of other structures such as blood vessels or nerves (see figure 1.3). Similar in function to a notch but appearing as a hole in a bone is a **foramen** (figure 1.5).

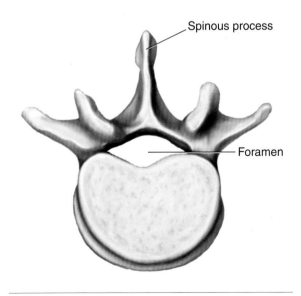

Figure 1.5 Superior (from above) view of a typical vertebra.

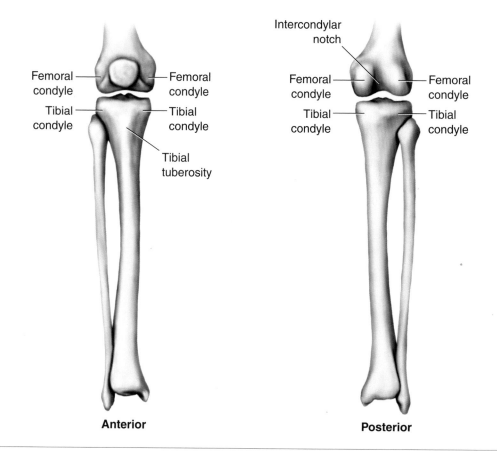

Figure 1.6 Anterior and posterior views of the knee.

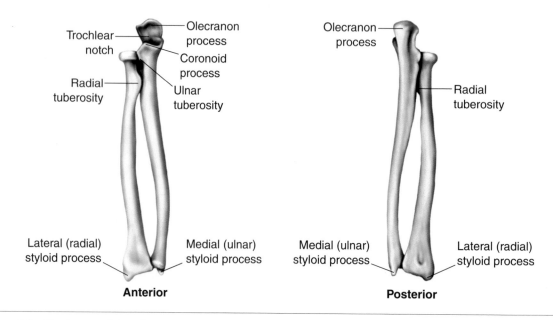

Figure 1.7 The bones of the elbow and their landmarks.

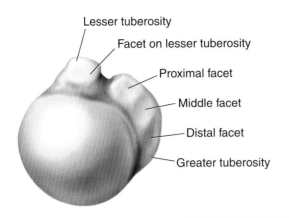

Figure 1.8 Superior view of the humerus.

Joints

The place where two or more bones join together anatomically is referred to as an **articulation.** The terms **joint** and *articulation* are interchangeable. Tying bones together at articulations are structures of dense, fibrous connective tissue known as **ligaments** (figure 1.9). A ligament is a cord, band, or sheet of strong, fibrous connective tissue that unites the articular ends of bones, ties them together, and facilitates or limits movements between the bones. Ligaments are not the sole support for the stability of joints. The muscles

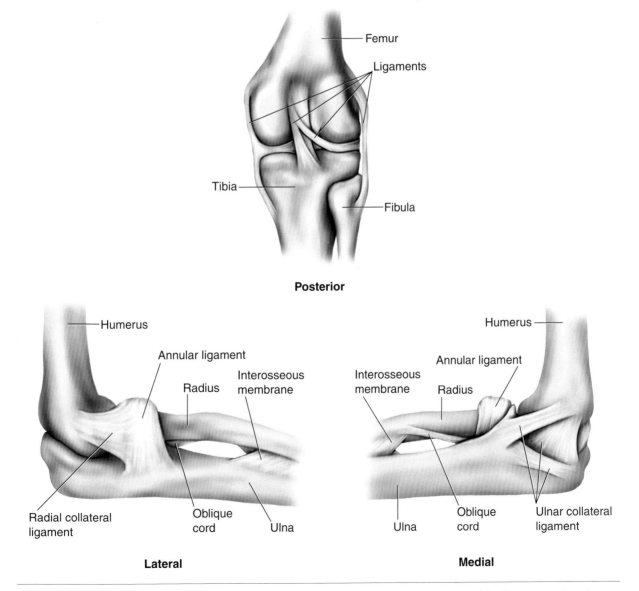

Figure 1.9 Medial and lateral views of the major ligaments of the elbow; posterior view of the ligaments of the knee.

that cross the joint and the actual formations of the articulating bones also contribute to joint stability.

There are two major forms of joints: diarthrodial and synarthrodial. **Diarthrodial joints** are distinguished by having a separation of the bones and the presence of a joint cavity. Diarthrodial joints are divided into six subdivisions by their shape (figure 1.10). The **hinge joint** has one concave surface, with the other surface looking like a spool of thread. The elbow joint is an example of a hinge type of diarthrodial joint. The **ball-and-socket** type of diarthrodial joint consists of the rounded head of one bone fitting into the cuplike cavity of another bone. Both the hip joint and the shoulder joint are examples of the ball-and-socket type of diarthrodial joint. The **irregular** type of diarthrodial joint consists of irregularly shaped surfaces that are typically either flat or slightly rounded. The joints between the bones of the wrist (carpals) are an example of this type of joint. Gliding movement occurs between the carpal bones. The **condyloid joint** consists of one convex surface fitting into a concave surface. Although the description of the condyloid joint is similar to that of the ball-and-socket joint, the difference is that the condyloid joint is capable of movement in only two planes about two axes, whereas the ball-and-socket joint is capable of movement in three planes about three axes. (Note: Planes and axes are discussed in chapter 2.) An example is the joint where the metacarpal bones of the hand meet the phalanges of the fingers. The **saddle joint** is often considered a modification of the condyloid joint. Both bones have a surface that is convex in one direction and concave in the opposite direction, like a saddle. These joints are rare, and the best example is the joint between the wrist and the thumb (carpometacarpal joint). In the **pivot joint,** one bone rotates about the other bone. The radius bone (of the forearm) rotating on the humerus (upper-arm bone) is an example of a pivot joint.

All of the diarthrodial joints are considered **synovial joints.** The synovial joints are where the greatest amount of movement occurs. They are characterized by a space between the articulating

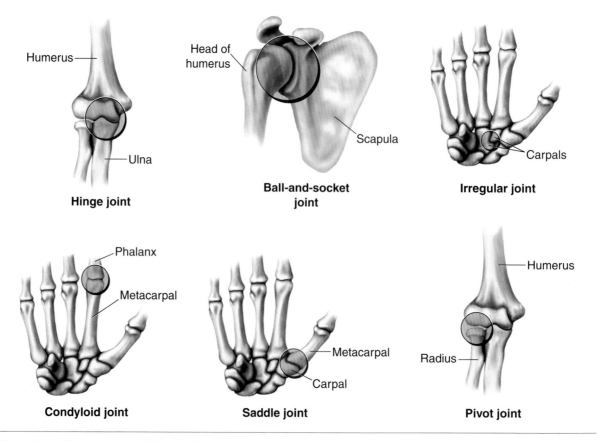

Figure 1.10 The six types of diarthrodial joints.

surfaces (figure 1.11); a synovial membrane lining the joint secretes synovial fluid for lubrication and provides nutrients to joint structures. Synovial joints have what is known as a joint (articular) **capsule.** Synovial joints are classified into four categories by the type of movement they permit in planes and about axes (figure 1.12). Joints between bones that allow only a gliding type of movement over each other are known as **nonaxial joints,** such as are found in the wrist and the foot. **Uniaxial joints,** such as the elbow joint, permit movement in only one plane about one axis. A **biaxial joint,** such as the wrist, permits movement in two planes, about two axes. The **triaxial joint** allows movement in three planes, about three axes, illustrated by the movements of the shoulder joint and the hip joint, which are both ball-and-socket joints.

Synarthrodial joints have no separation or joint cavity, unlike the diarthrodial joints. There are three subdivisions of synarthrodial joints (figure 1.13): sutured, cartilaginous, and ligamentous.

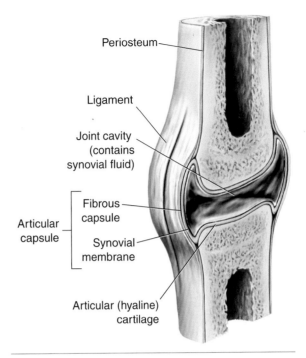

Figure 1.11 A diarthrodial (synovial) joint.

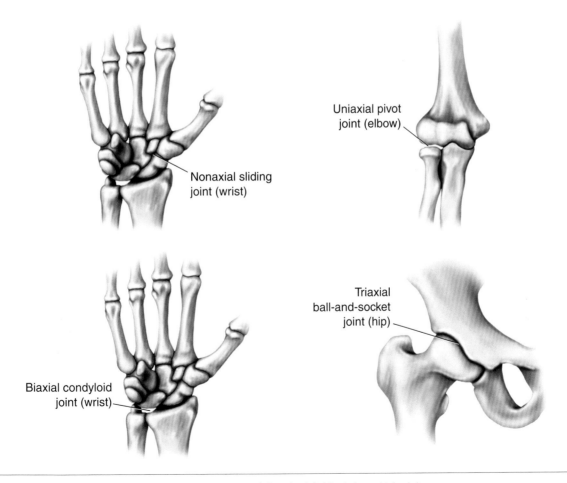

Figure 1.12 The four types of synovial joints: nonaxial, uniaxial, biaxial, and triaxial.

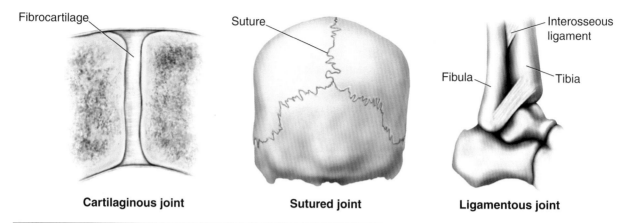

Fibrocartilage   Suture   Interosseous ligament

Fibula   Tibia

Cartilaginous joint   **Sutured joint**   **Ligamentous joint**

Figure 1.13   The three types of synarthrodial joints: cartilaginous, sutured, and ligamentous.

The **sutured joint** has no detectable movement and appears to be sewn (sutured) together like a seam in clothing. The bones of the skull are the classic examples of sutured joints. Because there is no movement in these joints, they are not discussed further in this text. **Cartilaginous joints** allow some movement, but cartilaginous joints other than those of the spinal column do not play a major role in movement. The cartilaginous joint contains **fibrocartilage** that deforms to allow movement between the bones and also acts as a shock absorber between them. The **ligamentous joints** tie together bones where there is very limited or no movement. The joints between two structures of the same bone (e.g., the coracoid process and acromion process of the scapula) and between the shafts of the forearm and lower-leg bones are examples of the ligamentous form of a synarthrodial joint.

Muscles

Muscle tissue is often categorized into three types: **smooth,** which occurs in various internal organs and vessels; **cardiac,** which is unique to the heart; and **skeletal,** which causes movement of the bones and their joints. For the purpose of looking at anatomy and movement, this text concentrates on the skeletal muscle. Skeletal muscle has the ability to stretch (extensibility), return to its original length when stretching ceases (elasticity), and shorten (contractility). The various forms of skeletal muscle are **fusiform, quadrate, triangular, unipennate, bipennate, longitudinal,** and **multipennate** fibers (figure 1.14). Most skeletal muscles are either fusiform or pennate fibers. Fusiform muscles are formed by long, parallel fibers and typically are involved in movement over a large range of motion. Pennate muscles consist of short, diagonal fibers

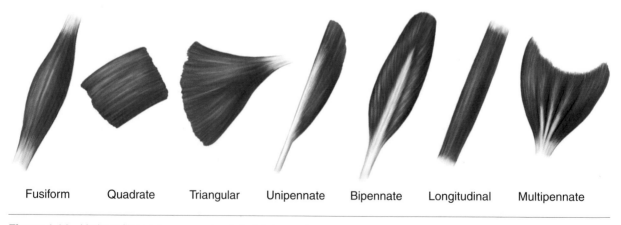

Fusiform   Quadrate   Triangular   Unipennate   Bipennate   Longitudinal   Multipennate

Figure 1.14   Various fiber arrangements of skeletal muscles.

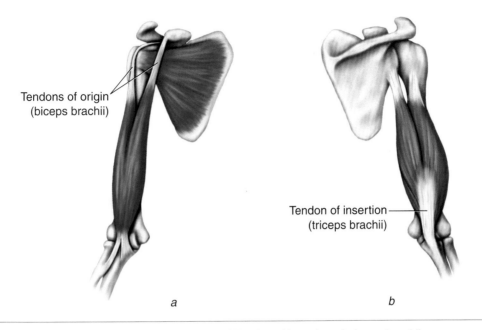

a
b

Figure 1.15 *(a)* Tendons of origin of biceps brachii. *(b)* Tendon of insertion of triceps brachii.

and are involved in movements that require great strength over a limited range of motion. The fibers of a muscle form the muscle belly. At either end of the belly of a muscle, a unique form of connective tissue, a **tendon,** attaches the muscle to the bones. Tendon is similar to ligament in that both are dense, regular connective tissue. The main difference is that the tendon tissue does not have as much elasticity as does the ligamentous tissue. Tendons are also extensible and elastic, like skeletal muscle, but they are *not* contractile. Tendons of skeletal muscles are usually defined as either **tendons of origin** or **tendons of insertion** (figure 1.15). Tendons of origin are usually longer and are attached to the proximal (closest to the center of the body) bone of a joint, which is typically the less mobile of the two bones of a joint. The tendons of insertion are shorter and are attached to the more distal (farther from the center of the body) bone of a joint, which is typically the more movable of the two bones of a joint. Additionally, because tendons cross bony areas or need to be confined to certain areas, the tendons are covered by connective tissue known as **tendon sheaths** (figure 1.16) to protect them from wear and tear from the bony structure they cross.

The general structure of skeletal muscle is shown in figure 1.17. Skeletal muscle is encased by a form of connective tissue known as the **epimysium.**

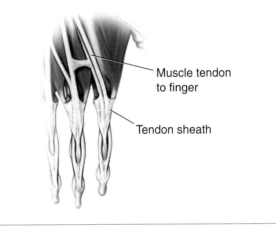

Figure 1.16 Tendon sheaths.

Within the epimysium are numerous **bundles** of **muscle fibers** that are individually wrapped in a fibrous sheath known as the **perimysium.** Within the perimysium are muscle fibers, which are in turn enclosed in a connective sheath known as the **endomysium.** A muscle fiber consists of a number of **myofibrils,** which are the contractile elements of muscle. Individual myofibrils are enclosed by a viscous material known as **sarcoplasm** and wrapped in a membrane known as the **sarcolemma.** Lengthwise, myofibrils consist of bands of alternating dark and light filaments of contractile protein known as **actin** and **myosin** (figures 1.17 and 1.18). This alternating pattern produces a

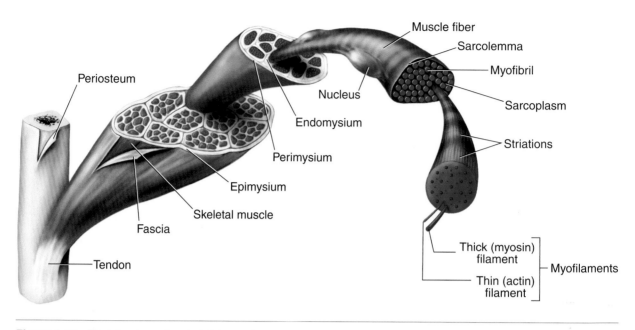

Figure 1.17   The structure of a skeletal muscle.

striped (striated) appearance when viewed under a microscope. A myofibril is divided into a series of **sarcomeres,** which are considered the functional units of skeletal muscle (figure 1.18). Sarcomeres contain an **I-band,** the light-colored portion where the protein filament actin occurs, and an **A-band,** the dark-colored area where the protein filament myosin occurs (figure 1.18). A sarcomere is that portion of a myofibril that appears between two

Z-lines (Z-lines bisect I-bands). Actin also occurs in A-bands. As the actin filaments extend into the A-band, they overlap with the myosin filaments contributing to the darker appearance at the edges of the A-band. The lighter-colored, central portion of the A-band is known as the **H-zone.** This region is lighter in color because actin does not extend into this area and because the myosin filament is thinner in the middle than at its outer edges. The

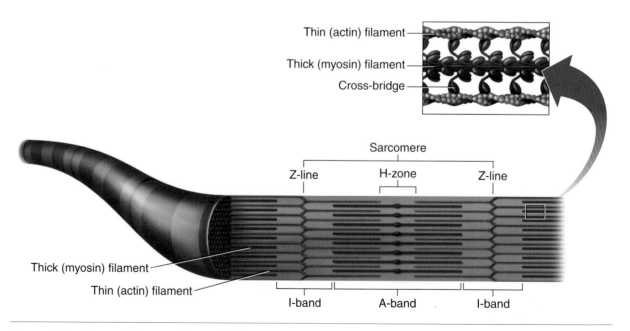

Figure 1.18   A muscle fiber and its myofibrils.

two protein filaments, actin and myosin, are the site of muscular movement (contraction). The myosin filament has **cross-bridges** (small extensions) that reach out at an angle toward the actin filaments (figure 1.18).

There are two primary types of skeletal muscle fibers, which are commonly known as **fast-twitch** and **slow-twitch fibers** (figure 1.19). Most muscles contain both types of fibers, but depending on heredity, function, and, to a lesser degree, training, some muscles contain more of one type of fiber than the other. Fast-twitch fibers are large and white and appear in muscles used to perform strength activities. The slow-twitch fibers are small and darker (red) than the fast-twitch fibers (primarily because they have a greater supply of hemoglobin). Slow-twitch fibers are slow to fatigue and are prevalent in muscles involved in performing endurance activities. A runner with a higher percentage of slow-twitch fibers in lower-extremity muscles is more likely to develop into a distance

Figure 1.19 Fast-twitch (light) and slow-twitch (dark) muscle fibers.

runner, whereas a runner with a higher percentage of fast-twitch fibers in lower-extremity muscles is more likely to become a sprinter.

Nerves

The body has three main nervous systems: an autonomic nervous system, a central nervous system, and a peripheral nervous system. The autonomic nervous system is involved with the glands and smooth muscle of the body. The central nervous system consists of the brain and spinal cord. The peripheral nervous system consists of 12 pairs of cranial nerves and 31 pairs of spinal nerves. The spinal nerves, divided in plexuses (networks of peripheral nerves), innervate (stimulate) the muscles to create movement. The major plexuses are the cervical, brachial, lumbar, sacral, and, to a limited extent, coccygeal (figure 1.20). Spinal column levels are typically referred to by specific vertebra. For example, C5 is the fifth cervical vertebra, T8 is the eighth thoracic vertebra, and L2 is the second lumbar vertebra.

The **nerve** (figure 1.21), or **neuron,** consists of a nerve cell body and projections from it, which are known as the **dendrite** and the **axon.** In a motor nerve, the dendrite receives information from surrounding tissue and conducts the nerve **impulse** *to* the nerve's **cell body** (responsible for neuron nutrition), and the axon conducts the nerve impulse *from* the cell body to the muscle fibers. Another structural component of a motor nerve is the **myelin sheath** that insulates the axon. The gaps in the myelin sheaths are known as the

---

Focus on . . . **MUSCLE VISCOSITY**

Viscosity is more easily understood if one considers motor oil used in an automobile. The viscosity (thickness) of the oil depends on the temperature: The oil either thins (as the temperature increases) or thickens (as the temperature decreases). Some authorities believe that one of the benefits of a warm-up before physical activity is that muscle viscosity changes with the increased temperature within the muscle tissue, which makes the muscle more able to endure the stress of the physical activity.

The study of muscle physiology and the physiology of exercise reveals other values of warming up. Anyone involved in prescribing physical activity (physicians, coaches, therapists, athletic trainers, and personal trainers) needs to understand the physiological factors involved in warm-up, including the effect on muscle viscosity.

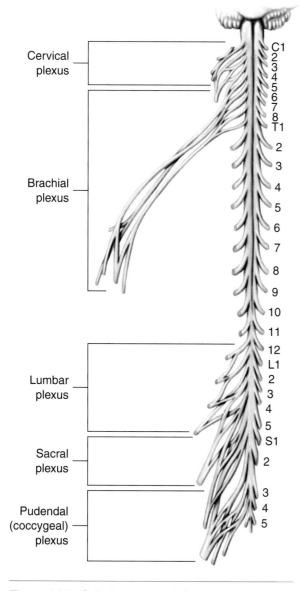

Figure 1.20   Spinal nerves and plexuses.

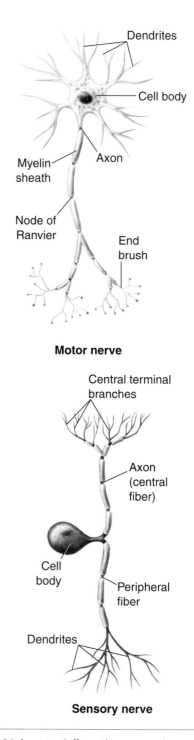

Motor nerve

Sensory nerve

Figure 1.21 A motor (efferent) nerve and a sensory (afferent) nerve.

nodes of Ranvier, which cause all impulses to "leap" between the myelin sheaths (from node to node), allowing the impulses to travel at higher speeds than they would across an unmyelinated sheathed axon. At the end of the axon is a structure known as the **motor end plate,** which consists of **end brushes** (terminal branches) that are in very close proximity to the muscle fibers. **Motor nerves** carry impulses *away* from the central nervous system, whereas **sensory nerves** (not discussed in this chapter) carry impulses *to* the central nervous system. Motor nerves are also referred to as **efferent nerves;** sensory nerves are also referred to as **afferent nerves.**

Blood Vessels

The **blood vessels** bring nutrients to the muscle tissue and carry away the waste products produced by the muscle tissues expending energy. When the heart pumps, blood moves out of the heart into a huge vascular tree consisting of arteries, arterioles (smaller arteries), capillaries, veins, and venules (small veins). There are three tissue layers (tunics) of the walls of arteries, veins, and capillaries (tunica intima, tunica media, and tunica adventitia). The middle layer (tunica media) contains various quantities of smooth muscle fibers depending upon the type of vessel. The arteries and arterioles (figure 1.22) distribute blood to the tissues, where capillaries provide the blood directly

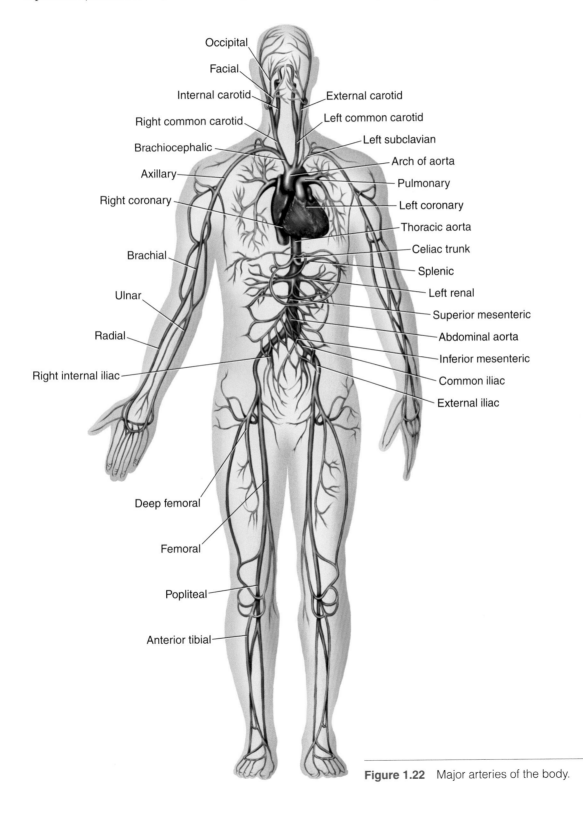

Figure 1.22 Major arteries of the body.

to the cells. The veins and venules (figure 1.23) collect the blood from the capillaries and return it to the heart. The middle wall of the arteries contains a vast amount of smooth muscle that contracts with the heart to pump the blood throughout the body. The veins contain small valves that permit blood to flow in only one direction (toward the heart). All three layers of tissue in the veins are much thinner as compared to the arteries. As a result, smooth muscle fibers are either absent or

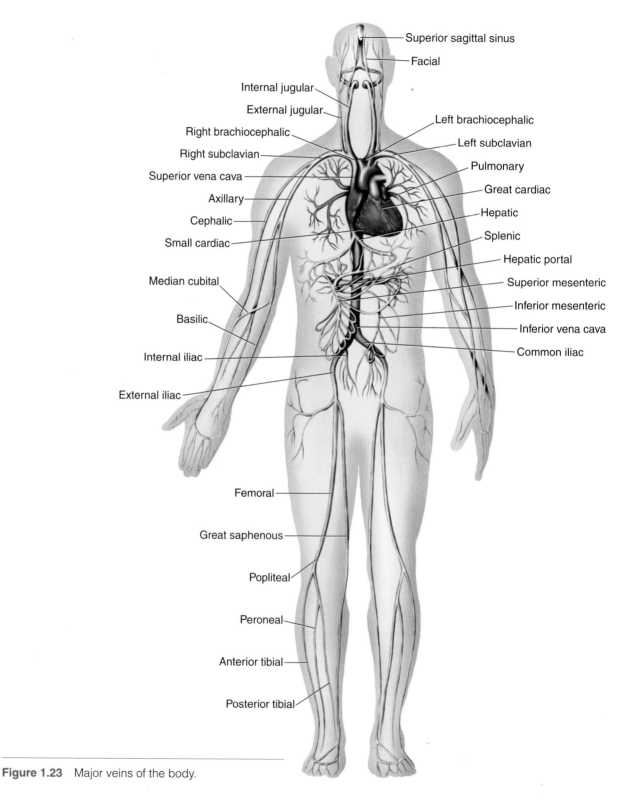

Figure 1.23   Major veins of the body.

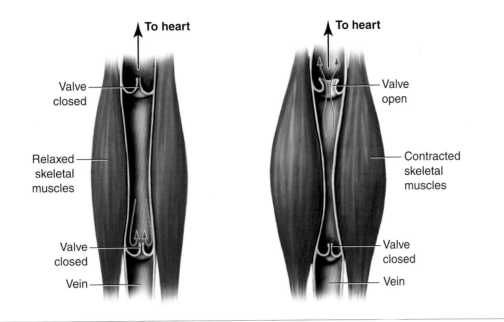

Figure 1.24 A vein's valve action.

are very few thin fibers in the tunica media of the veins. Although the veins possess either a minimal amount or no smooth muscle fibers in their walls compared to the arteries, they are assisted in returning blood to the heart by skeletal muscles as they contract and squeeze the veins between muscles and/or between muscles and bones (figure 1.24). The skeletal muscles act as muscular venous pumps that squeeze blood upward past each valve. Gravity also assists venous return in veins that are found above the heart. There are more valves in the veins of the extremities where upward blood flow is opposed by gravity.

Other Tissues

Other types of tissues associated with bones, joints, and muscles are fascia and bursa. **Fascia** is another form of fibrous connective tissue of the body that covers, connects, or supports other tissues. One form of fascia, the sarcolemma of muscle, has already been discussed. A **bursa** (figure 1.25) is a saclike structure that contains bursa fluid and protects muscle, tendon, ligament, and other tissues as they cross the bony prominences described earlier. The bursas provide lubricated surfaces to allow tendons to glide directly over bone without being worn away over time from friction.

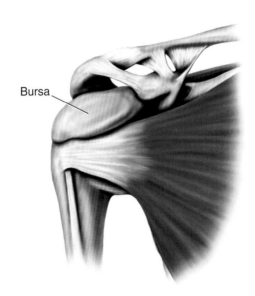

Figure 1.25 A typical bursa.

Motor Unit

We have now observed the bones, the ligaments that connect the bones to form articulations (joints), the muscles that cross joints and create movement, the nerves that innervate the muscles, and the blood vessels that supply all these structures. These structures are all considered essential to movement. We now look at the motor unit. Some textbooks examine the physiology of nerve

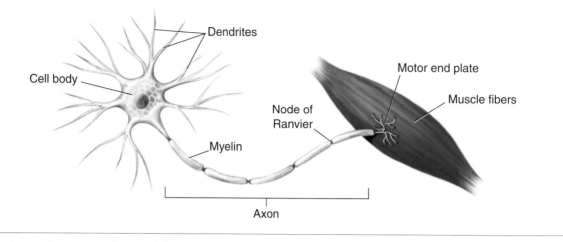

Figure 1.26   The motor unit, comprising a motor neuron and muscle fibers.

impulses that generate muscular contraction that, in turn, causes movement. In this book, we concentrate on the anatomy of the structures that actually produce movement.

A **motor unit** is defined as a motor nerve and all the muscle fibers it supplies (figures 1.26 and 1.27). The structural parts of the motor unit are the motor nerve and the muscle fiber. All the motor units together are referred to as the body's **neuromuscular system.**

The space between the end brushes of the motor end plate and the muscle fibers is known as the **myoneural junction.** Although the end brushes never actually come in direct contact with the muscle fibers, it is at the myoneural junction that the **synapse** (connection) is located. The synapse (figure 1.28) is the structure where the end brushes of the axon of a motor nerve release a chemical known as acetylcholine. This chemical stimulates the outer covering (sarcolemma) of the muscle fiber, which allows the impulse to continue to the muscle fiber and causes it to contract if the impulse is great enough to reach the muscle fiber's **threshold.** The space between the axon's end brush and the muscle's sarcolemma where acetylcholine passes from the nerve to the muscle tissue is known as the **synaptic cleft.** In summary, a nerve impulse travels from the spinal cord (or brain) to a dendrite of a spinal nerve, from the dendrite to the nerve's cell body, and from the cell body

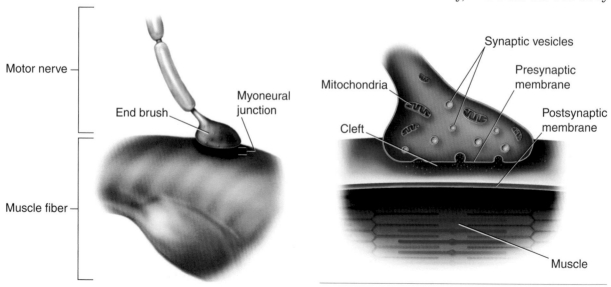

Figure 1.27   Schematic of a motor unit.

Figure 1.28   A neuromuscular synapse and its related structures.

over the nerve's axon to the axon's end brushes (motor end plate), where a chemical is released at the synapse.

Now that we have observed the anatomical structures responsible for the nerve impulse being transmitted to the muscle, let us look at the components that make up the other half of the motor unit, the muscle. As the impulse passes from nerve to muscle, **calcium** is released from the **sarcoplasmic reticulum** and **transverse tubules** (figure 1.29) within the muscle fibers, two structures closely allied with the actin and myosin protein filaments. The calcium release causes the myosin cross-bridges to wiggle or swivel in such a fashion that they contact the actin filaments surrounding them and cause the actin to move toward the center of the sarcomere (figure 1.30). This chemical communication within muscle fibers is responsible for the action known as the sliding filament mechanism.

In review, the impulse from the motor nerve crosses the synapse at the myoneural junction and activates the release of calcium through the sarcoplasmic reticulum and transverse tubules, causing the cross-bridges of the myosin protein filament to contact the actin protein filaments and produce movement of the actin filaments toward the center of the sarcomere, thus shortening the sarcomere.

Motor units differ widely in the number of muscle fibers innervated by one motor nerve. The ratio of muscle fibers per motor nerve can range from as low as 10 muscle fibers to as high as 2,000

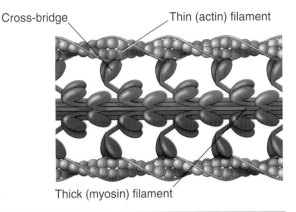

Figure 1.30 Cross-bridge formation and movement.

muscle fibers per one motor nerve. The lower the ratio of fibers to nerve, the more motor units are required to innervate all the fibers in a muscle. This is the case in muscles required to perform fine movements, such as those in the hand or the eye. By comparison, the biceps muscle performs elbow flexion and forearm supination, which are definitely not considered fine movements, and therefore has a very high ratio of muscle fibers to motor nerve.

If a stimulus from a nerve is intense enough to reach the threshold of a muscle fiber, all muscle fibers innervated by that nerve contract completely. There is no such thing as a partial contraction of a muscle fiber. This is referred to as the all-or-none theory of muscle contraction. Depending on the effort required (e.g., lifting a piece of paper versus a 50 lb [23 kg] weight), various degrees of muscle contraction (gradation of strength) are called on to perform the activity. The gradation of a muscular contraction depends on two major factors: the number of motor units recruited and the frequency with which they are stimulated. As the force required increases, more motor units are called into action. Additionally, they are stimulated more frequently. If impulses are sent rapidly enough to muscle fibers that they contract before totally relaxing from the previous contraction, a greater force of contraction can occur (up to a point). Once the muscle is receiving impulses at such a rate that it cannot relax, it reaches a state of continuous contraction known as tetanus (figure 1.31). The application of the term *tetanus* to this state of continuous contraction–the result of physical effort–should not be confused

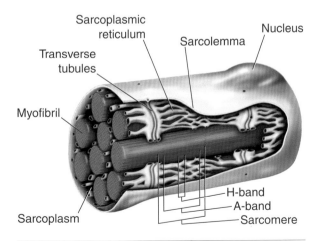

Figure 1.29 The sarcoplasmic reticulum and transverse tubules.

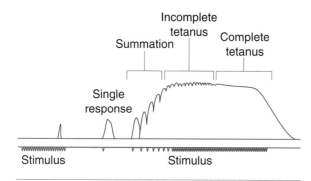

Figure 1.31   Muscular force increases with frequency of impulses.

with another usage of this term, by which is meant an infectious disease that can cause involuntary muscular contractions.

Courses in biology, human physiology, exercise physiology, kinesiology, biomechanics, and other fields examine the generation of a nerve impulse and the resulting contraction of muscle in detail. The preceding material should be considered an introductory overview and by no means a detailed analysis of the neuromuscular system and its motor units.

REVIEW OF TERMINOLOGY

The following terms were discussed in this chapter. Define or describe each term, and where appropriate, identify the location of the named structure either on your body or in an appropriate illustration.

A-band	fascia	perimysium
actin	fast-twitch fiber	periosteum
afferent nerve	fibrocartilage	pivot joint
anatomical landmark	foramen	process
articular cartilage	fossa	quadrate
articular surface	fusiform	saddle joint
articulation	hinge joint	sarcolemma
axon	H-zone	sarcomere
ball-and-socket joint	I-band	sarcoplasm
biaxial joint	impulse	sarcoplasmic reticulum
bipennate	irregular joint	sensory nerve
blood vessel	joint	skeletal muscle
bundle	ligament	slow-twitch fiber
bursa	ligamentous joint	smooth muscle
calcium	long bone	spine
capsule	longitudinal	spinous process
cardiac muscle	motor end plate	sutured joint
cartilaginous joint	motor nerve	synapse
cell body	motor unit	synaptic cleft
condyle	multipennate	synarthrodial joint
condyloid joint	muscle fiber	synovial joint
connective tissue	muscle tissue	tendon
cross-bridge	myelin sheath	tendon of insertion
dendrite	myofibril	tendon of origin
diaphysis	myoneural junction	tendon sheath
diarthrodial joint	myosin	threshold
efferent nerve	nerve	transverse tubule
end brush	nerve tissue	triangular
endomysium	neuroglia	triaxial joint
epicondyle	neuromuscular system	tubercle
epimysium	neuron	tuberosity
epiphyseal plate	node of Ranvier	uniaxial joint
epiphysis	nonaxial joint	unipennate
epithelial tissue	notch	Z-line
facet		

SUGGESTED LEARNING ACTIVITIES

1. Make a fist. List all the anatomical structures (starting with the brain) that were used for that action to occur.

2. Either at the dinner table or in a grocery store, look at the poultry and explain why a particular fowl (turkey, chicken) has meat of different colors in its various parts (legs, thighs, breasts, wings).
   a. Did the fowl's normal activities dictate more or less effort of certain body parts?
   b. What type of muscle fibers likely dominate the muscles of these various parts? Why?

3. From a standing position, rise up on your toes and stand that way for a few minutes (or as long as you can).
   a. What type of leg muscle (fast-twitch or slow-twitch) was primarily responsible for initially getting you into the toe-standing position?
   b. What type of leg muscle (fast-twitch or slow-twitch) was primarily responsible for sustaining you in the toe-standing position?

MULTIPLE-CHOICE QUESTIONS

1. A junction of two or more bones forming a joint is also known as
   a. an epiphysis
   b. a fossa
   c. an articulation
   d. a diaphysis

2. Which of the following terms does *not* appropriately fit with the other three?
   a. notch
   b. process
   c. tubercle
   d. tuberosity

3. The functional unit of skeletal muscle is known as
   a. a myofibril
   b. a sarcomere
   c. the A-band
   d. the I-band

4. A series of sarcomeres linked together is known as
   a. a myofibril
   b. a muscle
   c. actin
   d. myosin

5. The release of which of the following substances causes the cross-bridges to move, which in turn causes a sarcomere to shorten?
   a. actin
   b. myosin
   c. calcium
   d. sarcoplasm

6. A bundle of muscle fibers within a muscle is wrapped in a fibrous sheath known as the
   a. endomysium
   b. epimysium
   c. perimysium
   d. sarcolemma

FILL-IN-THE-BLANK QUESTIONS

1. Joints with no observable movement are known as _____ joints.

2. Saclike structures that protect soft tissues as they pass over bony projections are known as _____ _____.

3. A motor unit is a motor neuron and all the _____ _____ it supplies.

4. Dendrites conduct nerve impulses _____ _____ a cell body.

5. Axons conduct nerve impulses _____ _____ a cell body.

Movement

Now that you understand the structures involved in movement (bones, ligaments, muscles) and the terms used to describe them, let us look at the universal language that describes the movements performed by these structures.

When we describe a human movement, there is an anatomical "starting point" that is universally accepted as being the position all movements start from: the **anatomical position.** In this position, all joints are considered to be in a neutral position, or at 0°, with no movement having yet occurred. Occasionally, you might also hear the term **fundamental position.** Note carefully the only difference between the two positions (figure 2.1). The anatomical position is preferred to the fundamental position for any discussion of human movement because the hand position in the latter makes certain upper-extremity movements impossible. In the following sections, the description of any movement starts from the anatomical position.

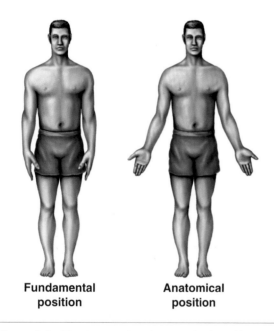

Fundamental position   **Anatomical position**

Figure 2.1   Fundamental and anatomical positions.

Anatomical Locations

Several terms are considered universal for discussing the spatial relationship between one anatomical structure and another. The term **superior** refers to something that is above or higher than another structure (e.g., your head is superior to your chest). The opposite term, **inferior,** means something is below or lower than another structure (e.g., your chest is inferior to your head). **Lateral** refers to something farther away from the midline of the body than another structure (e.g., your arms are lateral to your spinal column). **Medial** means a structure is closer to the midline of your body than another structure (e.g., your nose is medial to your ears). **Anterior** refers to a structure that is in front of another structure (e.g., your abdomen is anterior to your spinal column). **Posterior** refers to a structure that is behind another structure (e.g., your spinal column is posterior to your abdomen).

The terms **proximal** (close) and **distal** (far) are usually used in reference to structures of the extremities (arms and legs). Proximal means closer to the trunk, and distal means farther from the trunk (e.g., your knee is proximal to your ankle, and your hand is distal to your wrist). The term **dorsal** indicates the top side of a structure, such as the dorsal fin on the top of a fish. (The dorsal aspect of your hand is commonly called the back of your hand.) The term **volar** refers to the down side, or bottom aspect, of a structure. (The volar aspect of your wrist or hand is also referred to as the palmar aspect, whereas the volar aspect, or sole, of the foot is referred to as the **plantar** aspect.) Two terms refer to actions of the forearm and foot. The term **pronation** refers to the turning of the forearm from the anatomical position toward the body, resulting in the volar or palmar surface of the hand facing the body. Turning your foot downward and inward toward the other foot is referred to as pronation of the foot. **Supination,** the reverse of pronation, refers to turning the forearm outward from the pronated position and to the upward and outward movement of the foot away from the other foot.

Planes and Axes

Human movement that takes place from the starting (anatomical) position is described as taking place in a **plane** (a flat surface) about an **axis** (a straight line around which an object rotates). Movement is described from the anatomical position as the starting point. Muscles create movements of body segments in one or more of three planes that divide the body into different parts. These three specific planes are perpendicular (at right angles) to each other (figure 2.2). The **sagittal plane** (also known as the anteroposterior plane) passes from the front through the back of the body, creating a left side and a right side of the body. There could be any number of sagittal planes; however, there is only one **cardinal sagittal plane.** The term *cardinal* refers to the one plane that divides the body into equal segments, with exactly one half of the body on either side of the cardinal plane. Therefore, the cardinal sagittal plane divides the body into two equal halves on the left and right. The term *cardinal plane* appears in some texts as the *principal plane.* The terms are interchangeable.

The **horizontal plane** (also known as the transverse plane) passes through the body horizontally to create top and bottom segments of the body. There could be any number of horizontal planes but there is only one **cardinal horizontal plane,** which divides the body into equal top and bottom halves.

The **frontal plane** (also known as the lateral plane) passes from one side of the body to the other, creating a front side and a back side of the body. Again, there could be any number of frontal planes but there is only one **cardinal frontal plane,** which divides the body into equal front and back halves.

The point at the intersection of all three cardinal planes is the body's center of gravity. When all segments of the body are combined and the body is considered one solid structure in the anatomical position, the center of gravity lies approximately in the low back area of the spinal column. As body parts move from the anatomical position or weight shifts through weight gain, weight loss, or carrying loads, the center of gravity also shifts. No matter what the body's position or weight distribution, however, half of the weight of the body (and its load) will always be to the left and right, in front and behind, and above and below the center of gravity. The center of gravity of the body constantly changes with each movement, each change in weight distribution, or both.

Earlier we defined an axis as a straight line that an object rotates around. In the human body, we picture joints as axes and bones as the objects that rotate about them in a plane perpendicular to the axis. There are three main axes, and rotation is described as occurring in a plane about the axis that is perpendicular to the plane (figure 2.2). The sagittal plane rotates about a **frontal horizontal axis** (figure 2.2*a*).

Hands on . . . The knee joint is a frontal horizontal axis, and the lower leg is the object that moves in the sagittal plane when you bend your knee.

The horizontal plane rotates about a **vertical** (longitudinal) **axis** (figure 2.2*b*).

Hands on . . . As you turn your head to the left and right as if to silently say no, your head rotates in a horizontal plane about the vertical axis created by your spinal column.

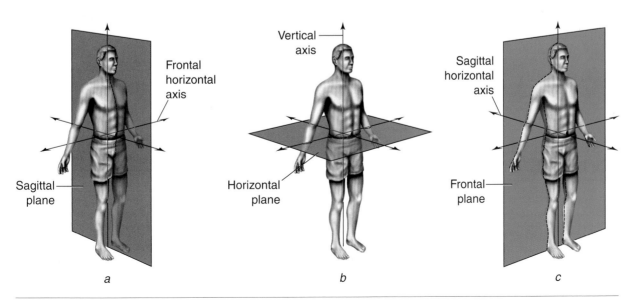

Figure 2.2 The cardinal planes and axes.

The frontal plane rotates about the **sagittal horizontal axis** (figure 2.2*c*).

Hands on . . . When you raise your arm to the side, your shoulder joint is the sagittal horizontal axis and your arm is the object moving in the frontal plane.

For a summary of the relationship of anatomical planes and associated axes, see table 2.1.

Table 2.1   Planes, Axes, and Fundamental Movements

Plane	Axis	Movements
Sagittal (anteroposterior)	Frontal horizontal	Flexion and extension
Frontal (lateral)	Sagittal horizontal	Abduction and adduction
Horizontal (transverse)	Vertical	Rotation

Fundamental Movements

Again, be reminded that movement takes place in a plane about an axis. There are three planes and three axes with two **fundamental movements** possible in each plane. In the sagittal plane, the fundamental movements known as **flexion** and **extension** are possible. Flexion is defined as the decreasing of the angle formed by the bones of the joint (figure 2.3). In flexion of the elbow joint, the angle between the forearm and upper arm decreases. Extension is defined as the increasing of the joint angle (figure 2.4). Returning a joint in flexion to the anatomical position is considered extension. Fundamental movements in the frontal plane are known as **abduction** and **adduction**. Abduction is defined as movement away from the midline of the body (figure 2.5). As you move your arm away from the side of your body in the frontal plane, you are abducting the shoulder joint. Movement toward the midline of the body is defined as adduction (figure 2.6). Returning your arm from an abducted shoulder position to the anatomical position is adduction. The fundamental movement in the horizontal plane is simply defined as **rotation** (figure 2.7). The earlier example of shaking your head no is rotation of the head. For describing movement in the upper (arm) and

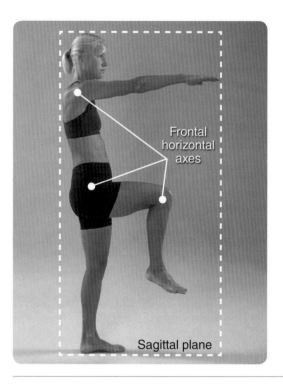

Figure 2.3   Flexion at the shoulder, hip, and knee.

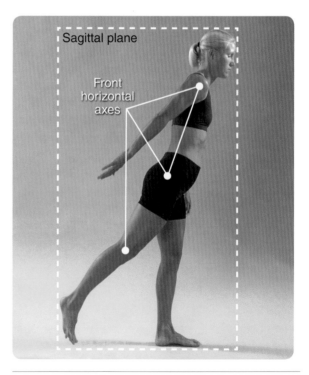

Figure 2.4   Extension at the shoulder, hip, and knee.

lower (leg) extremities, the terms **external rotation** and **internal rotation** are often used (figure 2.8). When the anterior (front) surface of the arm or leg rotates laterally (away from the midline of

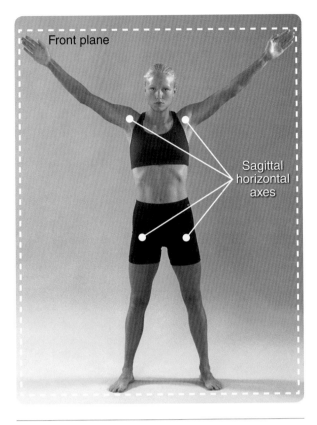

Figure 2.5   Abduction at the shoulder and hip.

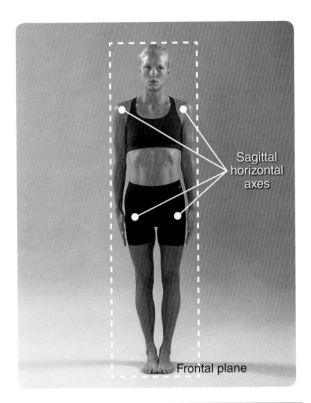

Figure 2.6   Adduction at the shoulder and hip as a return from the abducted position in figure 2.5.

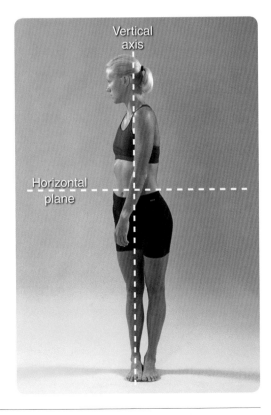

Figure 2.7   Rotation as a twisting movement of the spine.

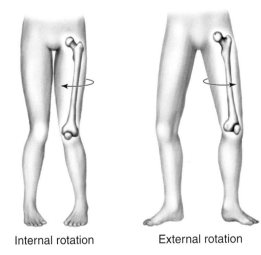

Internal rotation   External rotation

Figure 2.8   Internal rotation and external rotation of the lower extremity.

the body), this is defined as external rotation (or lateral rotation). When the anterior surface of the arm or leg rotates medially (toward the midline of the body), this is defined as internal rotation (or medial rotation).

Joints capable of creating movement in two (biaxial) or three (triaxial) planes are also capable of another movement, **circumduction**, which, because it combines two or more fundamental movements, is *not* considered a fundamental movement of any joint. When movement occurs in the two or three planes in a sequential order, the joint is said to be circumducting. Moving your arm at the shoulder joint in a "windmill" motion is an example of circumduction (figure 2.9).

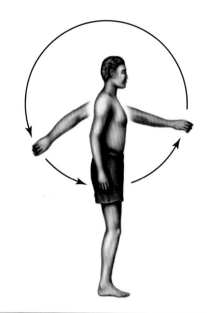

Figure 2.9   Circumduction at the shoulder.

REVIEW OF TERMINOLOGY

The following terms were discussed in this chapter. Define or describe each term and, where appropriate, identify the location of the named structure either on your body or in an appropriate illustration.

abduction
adduction
anatomical position
anterior
axis
cardinal frontal plane
cardinal horizontal plane
cardinal sagittal plane
circumduction
distal
dorsal
extension

external rotation
flexion
frontal horizontal axis
frontal plane
fundamental movement
fundamental position
horizontal plane
inferior
internal rotation
lateral
medial
plane

plantar
posterior
pronation
proximal
rotation
sagittal horizontal axis
sagittal plane
superior
supination
vertical axis
volar

SUGGESTED LEARNING ACTIVITIES

1. Stand in the anatomical position.
   a. Flex your knee joint. In what plane did movement occur? About what axis did movement occur?
   b. Abduct your hip joint. In what plane did movement occur? About what axis did movement occur?
   c. Rotate your head so your chin touches your left shoulder. In what plane did movement occur? About what axis did movement occur?

2. The body's center of gravity is considered the point where the three cardinal planes of the body intersect.
   a. In what direction would your center of gravity shift if you were to put a backpack full of books on your shoulders?
   b. In what direction would your center of gravity shift if you were to carry a large book in both hands in front of you below your waist?
   c. In what direction would your center of gravity shift if you were to carry a heavy briefcase in your left hand with your arm fully extended at your side?

MULTIPLE-CHOICE QUESTIONS

1. Which of the following movements is defined as movement in the frontal plane toward the midline of the body?
   - a. abduction
   - b. adduction
   - c. flexion
   - d. extension

2. For movement of the shoulder joint to occur in the frontal plane, which of the following joint actions must take place?
   - a. internal rotation
   - b. circumduction
   - c. flexion
   - d. abduction

3. A motion occurring in the horizontal (transverse) plane about a vertical axis is known as
   - a. abduction
   - b. rotation
   - c. adduction
   - d. extension

4. Movement taking place in the frontal plane is about the
   - a. frontal horizontal axis
   - b. sagittal horizontal axis
   - c. vertical axis
   - d. horizontal axis

FILL-IN-THE-BLANK QUESTIONS

1. When the angle formed at a joint diminishes and the movement takes place in the sagittal plane, the movement is known as _____.

2. Joint motion is typically described as taking place about an axis and within _____.

3. An axis of the body that passes through the body horizontally from front to back is known as a _____ _____ axis.

4. When referring to a structure of an extremity being closer to the trunk than another structure, we say that it is _____ to the other structure.

PART II

Upper Extremity

The Shoulder

CHAPTER OUTLINE

Any discussion of the shoulder must start with the fact that the shoulder is actually two distinct anatomical structures: the **shoulder girdle** and the **shoulder joint.** The shoulder girdle consists of the **clavicle** and **scapula** bones, whereas the shoulder joint is formed by the scapula and the **humerus** bones. The primary function of the shoulder girdle is to position itself to accommodate movements of the shoulder joint.

Bones of the Shoulder Girdle

Two bones make up the structure known as the shoulder girdle: the clavicle and the scapula (figure 3.1). The clavicle is a long, slender, S-shaped bone that attaches to the **sternum** (breastbone) at the medial end and to the scapula at the lateral end. The clavicle is often referred to as the collarbone. It is the only bony attachment that the upper extremity has to the trunk. Because of its shape, the fact that it is held in place at either end by strong, unyielding ligaments, and that fact that it has little protection from external forces (with the exception of skin), the clavicle is an often-fractured bone.

The lateral end of the clavicle is referred to as the **acromial end,** and the medial end is referred to as the **sternal end** (figure 3.2). Prominent bony landmarks are observed on the superior and inferior views (figure 3.3) and include the **deltoid tubercle,** the **conoid tubercle,** the **trapezoid line,** the **costal tuberosity,** and the **subclavian groove.** These structures are important as places of attachment for soft tissue.

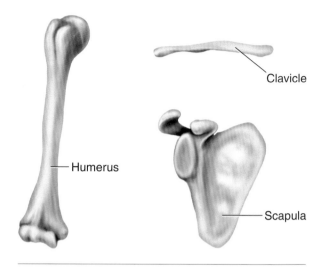

Figure 3.1   The bones of the shoulder, anterior view.

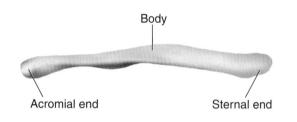

Figure 3.2   Sternal and acromial ends of the clavicle.

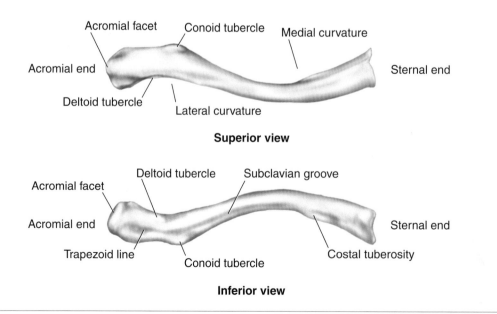

Superior view

Inferior view

Figure 3.3   Prominent bony landmarks of the clavicle, superior and inferior views.

The scapula is the large, triangular, winglike bone in the upper posterior portion of the trunk. This bone, sometimes referred to as the shoulder blade, is considered a bone of both the shoulder girdle and the shoulder joint. Figure 3.4 illustrates the many bony prominences of the scapula, including the **lateral** and **medial borders** and the **inferior angle** at the junction of the two borders.

Hands on . . . You can locate these structures on a partner (figure 3.5).

Additionally, the **superior border** at the medial end of the scapula becomes the **superior angle** and at its lateral end has a notch known as the **scapular notch.** Two large bony prominences at the superior lateral portion of the scapula are known as the **coracoid process** (anterior) and the **acromion process** (posterior).

Hands on . . . Either on yourself or a partner, palpate (touch) both the coracoid and acromion processes (figures 3.6 and 3.7).

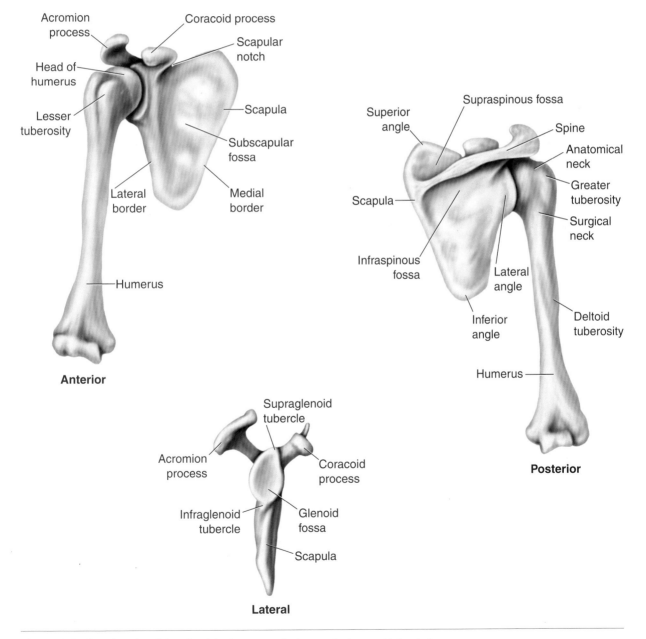

Figure 3.4 Landmarks of the shoulder bones; anterior, posterior, and lateral views.

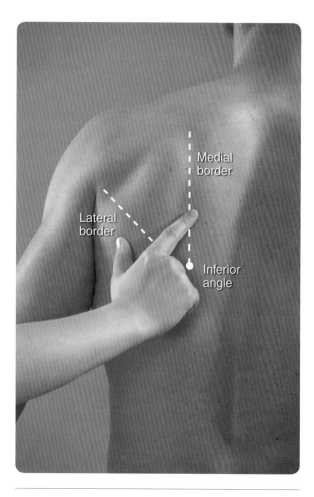

Figure 3.5 Identifying the lateral border, medial border, and inferior angle of the scapula.

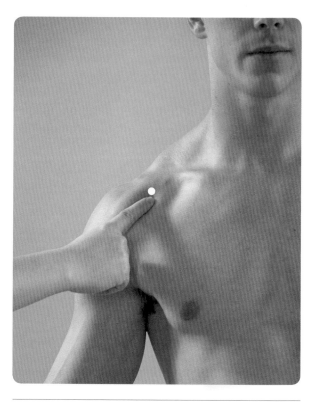

Figure 3.6 Locating the coracoid process.

The acromion process is the lateral expansion of a ridge of bone approximately one third of the way down the posterior aspect of the scapula. This ridge of bone is known as the **spine** of the scapula.

Hands on . . . Apply pressure with your index and middle fingers along the upper third of your partner's scapula and you'll feel this spine (figure 3.8).

Most laterally, the scapula forms a smooth, round, slightly depressed surface known as the **glenoid fossa.** This cavity forms the socket for the shoulder joint. Above and below the glenoid fossa are two bony prominences known, respectively, as the **supraglenoid** and **infraglenoid tubercles.** The smooth area of bone between the lateral and medial borders of the scapula on the anterior sur-

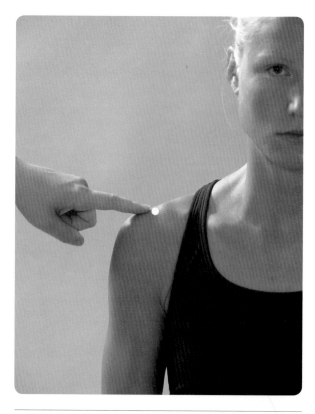

Figure 3.7 Locating the acromion process.

although somewhat shallow, is considered the socket of the joint.

The "ball" of the shoulder joint is the structure known as the **head** of the humerus. This chapter discusses the humerus only because it is involved with the shoulder joint (the proximal end; see figure 3.9). The humerus at its distal end is discussed further in the chapter on the elbow joint.

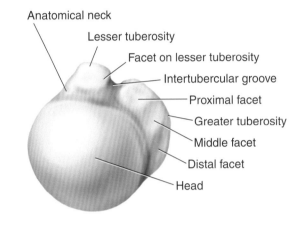

Figure 3.9 Landmarks of the proximal end of the humerus.

The head of the humerus is separated from the shaft of the bone by two necks. The **anatomical neck** is located between the head of the humerus and two bony prominences known as the **greater** and **lesser tuberosities**. The **surgical neck** (see figure 3.4) is actually the upper portion of the shaft of the humerus. A groove known as the intertubercular (bicipital) groove is created by the greater and lesser tuberosities. Atop both the lesser and greater tuberosities appear four flat surfaces known as facets. The lesser tuberosity has one facet, whereas the greater tuberosity has three: a **proximal, middle,** and **distal facet.** Approximately halfway down the shaft of the humerus, on the lateral surface, is a bony prominence known as the **deltoid tuberosity** (see figure 3.4).

Hands on . . . Apply pressure to either your own or your partner's humerus approximately halfway between the head and the distal end at the elbow on the lateral aspect; you'll feel the bump of the deltoid tuberosity (figure 3.10).

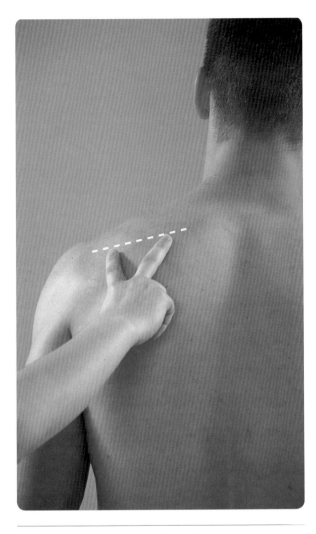

Figure 3.8 Locating the spine of the scapula.

face is known as the **subscapular fossa.** On the posterior surface of the scapula, the smooth bony surfaces above and below the spine are known, respectively, as the **supraspinous fossa** and **infraspinous fossa** (see figure 3.4).

Bones of the Shoulder Joint

The shoulder joint is the articulation between the scapula and the humerus (bone of the upper arm). The joint is known as the **glenohumeral (GH) joint** because of the two articulating bony surfaces. The prominent structure of the scapula regarding the shoulder joint is the anatomical area labeled the glenoid fossa. The shoulder joint is classified as a ball-and-socket joint, and the glenoid fossa,

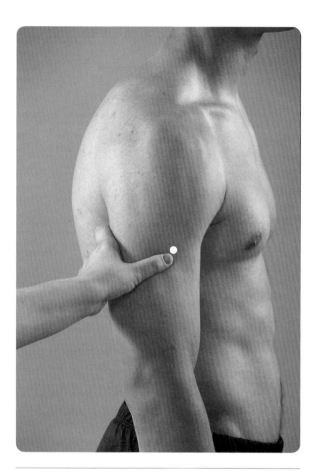

Figure 3.10   Locating the deltoid tuberosity.

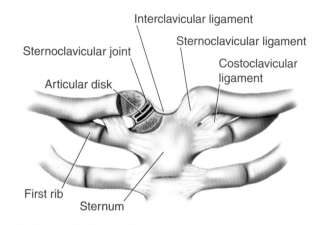

Figure 3.11   Sternoclavicular, costoclavicular, and interclavicular ligaments and the articular disc.

Joints and Ligaments of the Shoulder Girdle

The shoulder girdle has two joints, one at either end of the clavicle, known as the **acromioclavicular (AC)** and **sternoclavicular (SC) joints.** Movement in the SC joint is slight in all directions and of a gliding, rotational type. The joint receives its stability both from its bony arrangement, because the sternal end of the clavicle lies in the **clavicular notch** of the manubrium of the sternum (see chapter 8), and from the ligamentous arrangement that ties the clavicle and sternum together.

Three primary ligaments are responsible for the SC articulation (figure 3.11). The **sternoclavicular ligament,** with **anterior, superior,** and **posterior fibers,** and two other ligaments help to stabilize the SC articulation: the **costoclavicular ligament,** which secures the sternal end of the clavicle to the first rib, and the **interclavicular ligament,** which

secures the sternal ends of both clavicles into the clavicular notch of the manubrium of the sternum. Also, an **articular disc** is present between the sternal end of the clavicle and the clavicular notch of the manubrium of the sternum.

Hands on . . . Place your index finger in the space between each of your clavicles; you will feel the clavicular notch formed between both SC joints (figure 3.12).

Figure 3.12   Locating the sternoclavicular joint.

The joint between the lateral (acromial) end of the clavicle and the scapula is divided into two separate areas: the AC joint and the **coracoclavicular joint.** Simply described, the lateral end of the clavicle articulates with both the acromion process and the coracoid process of the scapula (figure 3.13).

Hands on . . . Palpate the area at the lateral end of the clavicle (figure 3.14). You should both feel and see the bump that is the AC joint.

The AC joint is the articulation between the acromion process of the scapula and the acromial end of the clavicle. There is a slight gliding type of movement between the two bones of this joint when elevation and depression of the acromial end of the clavicle and the acromion process of the scapula take place. The **acromioclavicular ligament** functions as the joint capsule, tying together and totally surrounding the lateral end of the clavicle and the acromion process of the scapula.

The other shoulder girdle joint, the coracoclavicular joint, is sometimes considered a component of the acromioclavicular joint and

Figure 3.14 Locating the acromioclavicular joint.

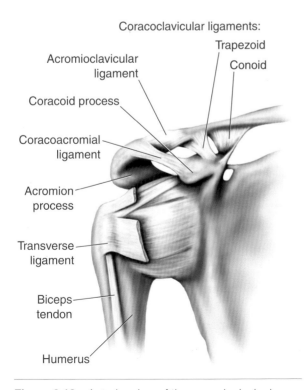

Figure 3.13 Anterior view of the acromioclavicular and coracoclavicular ligaments.

sometimes treated as a separate joint. The joint is the articulation between the lateral (acromial) end of the clavicle and the coracoid process of the scapula. Two ligaments run between the coracoid process of the scapula and the inferior surface of the clavicle. The two ligaments, the conoid and the trapezoid, are often referred to as a single ligament, the **coracoclavicular ligament.** Although some people do not consider the coracoclavicular joint a true joint, slight movement occurs in all directions in the articulation. The **trapezoid ligament** is the more lateral component of the coracoclavicular ligament and runs from the superior aspect of the coracoid process of the scapula to the anterior inferior aspect of the clavicle. It functions to oppose forward, upward, and lateral movement of the lateral aspect of the clavicle. The **conoid ligament** is the medial component of the coracoclavicular ligament and runs from the superior aspect of the coracoid process of the scapula

Focus on . . . **SHOULDER SEPARATION**

A sprain (partial or complete tearing) of the acromioclavicular and coracoclavicular ligaments results in a visible gap between the clavicle and the scapula, a classic illustration of what is known as a shoulder separation (figure 3.15). The separation is actually a widening of the space between the lateral end of the clavicle and the acromion process of the scapula. The weight of the upper extremity often reveals this gap when the affected shoulder girdle is compared with the unaffected one. When you hear the term *shoulder separation,* know that it more correctly refers to the acromioclavicular joint of the shoulder girdle.

Falling out of a tree, being tackled in a football game, being taken down to the mat in a wrestling match, or slipping on ice and falling on one's shoulder can all lead to spraining the acromioclavicular joint. To further study the topic of sprains, consult athletic training or sports medicine texts and talk with athletic trainers and physicians who specialize in sports medicine. Knowledge of the basic anatomy of the shoulder will enhance your ability to understand the literature and communicate with these specialists.

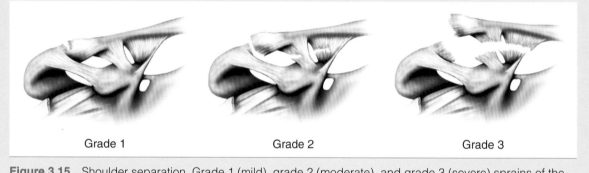

Grade 1   Grade 2   Grade 3

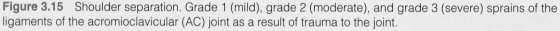

Figure 3.15   Shoulder separation. Grade 1 (mild), grade 2 (moderate), and grade 3 (severe) sprains of the ligaments of the acromioclavicular (AC) joint as a result of trauma to the joint.

to the posterior inferior aspect of the clavicle. It functions to oppose backward, upward, and medial movement of the lateral aspect of the clavicle. The coracoclavicular ligament is a strong supporter of the acromioclavicular ligament. Loss of these ligaments results in separation of the upper extremity from the trunk of the body.

Ligaments of the Shoulder Joint

The shoulder joint is the articulation between the head of the humerus and the glenoid fossa of the scapula. The ligaments of the shoulder joint (figure 3.16) include the **capsular ligament,** the **glenohumeral ligament** (superior, inferior, and middle sections), and the **coracohumeral ligament.** The cap-

sular ligament attaches the anatomical neck of the humerus and the circumference of the glenoid of the scapula. The glenohumeral ligaments are located beneath the anterior surface of the joint capsule and reinforce the capsule. The **superior glenohumeral ligament** runs between the upper surface of the lesser tuberosity of the humerus and the superior edge of the glenoid of the scapula. The **middle glenohumeral ligament** runs between the anterior surface of the lesser tuberosity of the humerus and the anterior edge of the glenoid of the scapula. The **inferior glenohumeral ligament** runs between the lower anterior surface of the lesser tuberosity of the humerus and the lower anterior edge of the glenoid of the scapula. The coracohumeral ligament runs between the anatomical neck of the humerus, near the greater tuberosity, and the lateral aspect of the coracoid process of the scapula.

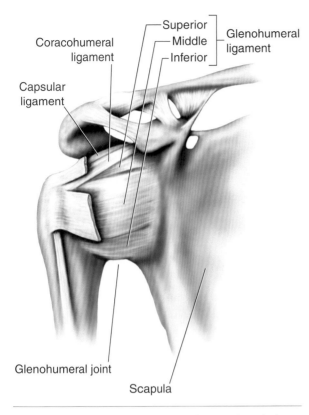

Figure 3.16   The capsular, coracohumeral, and glenohumeral ligaments.

Other Ligaments of the Shoulder

In addition to ligaments of the shoulder joint and girdle, other ligaments of the shoulder include those specific to the scapula and the humerus.

Ligaments of the Scapula

Although we typically think of ligaments as tying bones together to form articulations, some ligaments run from one aspect of a bone to another aspect of the same bone, serving some other function than forming joints. Four such ligaments are prominent on the scapula (figure 3.17). The **superior transverse scapular ligament** crosses the scapular notch, converting the notch into a foramen through which the suprascapular nerve passes. The **inferior transverse scapular ligament** crosses from one edge to the other of the **great scapular notch.** This ligament forms a tunnel for the passage of the suprascapular nerve that innervates (stimulates), and the

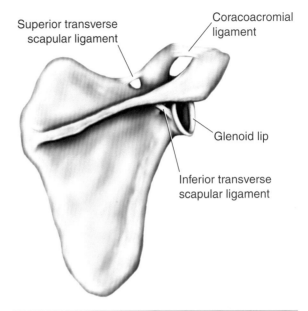

Figure 3.17   The superior and inferior transverse scapular ligaments, the coracoacromial ligament, and the glenoid lip.

transverse scapular blood vessels that supply blood to, the **infraspinatus** muscle. The **coracoacromial ligament** crosses between the coracoid process and the acromion process of the scapula. Although a ligament of the scapula, this ligament functions to limit superior movement of the humeral head. The **glenoid lip** (also known as the **glenoid labrum**) (figure 3.18) is a ligament that forms an edge

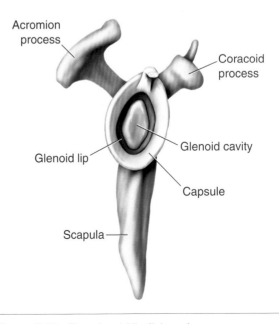

Figure 3.18   The glenoid lip (labrum).

around the entire circumference of the glenoid of the scapula. The glenoid lip helps deepen the glenoid fossa for the head of the humerus to add to the stability of the shoulder joint.

Ligament of the Humerus

On the anterior surface of the proximal end of the humerus, two structures were discussed: the greater and lesser tuberosities. Also mentioned was the **intertubercular groove** formed between these structures. Crossing the intertubercular groove is a ligament known as the **transverse humeral ligament** (figure 3.19). This ligament has one function: to hold the tendon of origin of the long head of the biceps brachii muscle in the groove.

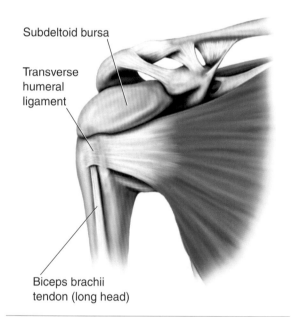

Subdeltoid bursa

Transverse humeral ligament

Biceps brachii tendon (long head)

Figure 3.19 Transverse humeral ligament.

Fundamental Movements and Muscles of the Shoulder Girdle

In this chapter and all succeeding chapters, the fundamental movements involving the anatomical area under discussion are introduced first. A discussion of the muscles involved in those movements is then presented.

Movements of the Shoulder Girdle

Movements of the shoulder girdle are primarily for the purpose of accommodating shoulder joint movement through changing positions of the glenoid of the scapula. Although it was stated earlier that fundamental movements are those confined to a single plane about a single axis, the shoulder girdle is an exception. Remember that the starting position for description of all fundamental movements of any joint is the anatomical position. There are four fundamental movements of the shoulder girdle: **elevation, depression, abduction,** and **adduction.** Because of the relationships between the clavicle and scapula and between the scapula and the thorax, on which it is positioned, movement exclusively in one plane about one axis is not always possible. Shoulder girdle fundamental movements are described relative to the direction in which the scapula moves.

Elevation of the shoulder girdle is defined as a superior (upward) movement of the scapula in the frontal plane (figure 3.20).

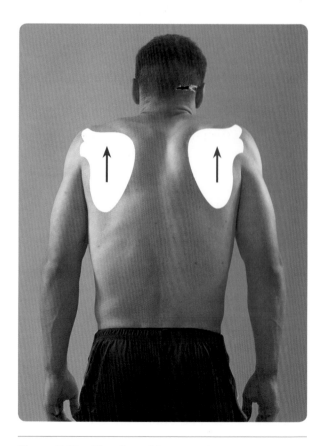

Figure 3.20 Elevation of the scapulas.

Depression of the shoulder girdle may be described as inferior (downward) movement of the scapula in the frontal plane (figure 3.21) but should more correctly be described as return from elevation. Because of the anatomical starting position, depression of the shoulder girdle is not possible. Eccentric contraction (lengthening) of muscles that concentrically contracted (shortened) to cause elevation of the shoulder girdle results in depression of the shoulder girdle (return to the anatomical position).

Abduction of the shoulder girdle cannot be simply defined as movement away from the midline of the body by the scapula. Because the scapula is tied to the clavicle (at the AC joint) by ligaments and to the chest **(thorax)** by muscle tissue, the scapula cannot move purely laterally away from the midline of the body. The scapula must rotate about the distal end of the clavicle (at the AC joint) and tilt as it glides along the chest (thorax). Thus, abduction of the shoulder girdle is more correctly defined as **upward rotation** and **lateral tilt** of the scapula

(figure 3.22). The upward rotation is defined as the upward movement of the glenoid of the scapula that accommodates shoulder joint movement. Additionally, the scapula tilts laterally as it glides along the curvature of the chest (thorax). This movement, when performed by both scapulas, is also referred to as **protraction.** Hugging another person by placing your arms around that person requires you to protract both of your shoulder girdles.

Adduction of the shoulder girdle cannot be simply defined as movement toward the midline of the body by the scapula. Again, because the scapula is tied to the clavicle (at the AC joint) by ligaments and to the chest (thorax) by muscle tissue, the scapula cannot move purely medially toward the midline of the body. The scapula rotates about the distal end of the clavicle (at the AC joint) and tilts as it glides along the chest (thorax). Thus, adduction of the shoulder girdle is more correctly defined as **downward rotation** and **medial tilt** of the scapula (figure 3.23). The downward rotation is defined as the downward movement of the glenoid of the

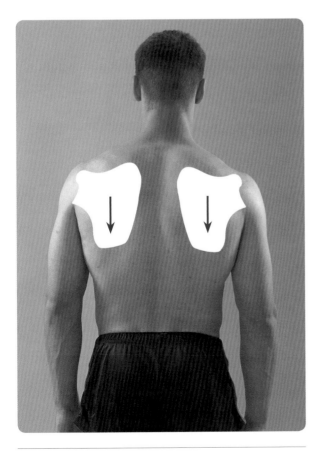

Figure 3.21 Depression of the scapulas.

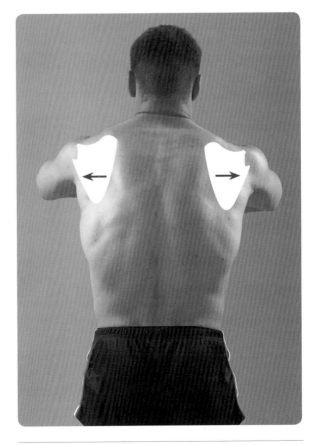

Figure 3.22 Abduction of the scapulas.

scapula that accommodates shoulder joint movement. Additionally, the scapula tilts medially as it glides along the curvature of the chest (thorax). This movement, when performed by both scapulas, is also referred to as **retraction.** Bringing your shoulders back, as when standing "at attention," requires you to retract both of your shoulder girdles.

Anterior Muscles of the Shoulder Girdle

Six muscles are primarily involved in producing the fundamental movements of the shoulder girdle. Three muscles are anatomically anterior to the shoulder girdle bones, and three are posterior to it. The anterior muscles of the shoulder girdle include the pectoralis minor, serratus anterior, and subclavian muscle.

> **Pectoralis minor:** The pectoralis minor originates from the third, fourth, and fifth ribs and inserts on the coracoid process of the scapula (figure 3.24). Because the ribs are the more stable attachment, contraction of the pectoralis minor results in the coracoid process of the scapula being pulled toward the ribs (shoulder girdle downward rotation, or adduction).

> **Serratus anterior:** Originating from the anterior lateral aspects of the upper nine ribs, the serratus anterior inserts on the anterior surface of the vertebral (medial) border of the scapula (figure 3.24). Because the more stable attachment of the serratus anterior is on the ribs, contraction of the muscle results in abduction (upward rotation and lateral tilt) of the shoulder girdle.

Hands on . . . Have your partner raise his or her arm above the head, with the hand on the back of the head (figure 3.25). The serratus anterior should be prominently displayed laterally along the ribs.

> **Subclavian muscle:** The subclavian muscle originates on the first rib and inserts on the subclavian groove of the clavicle. Its main function is to assist the ligaments of the SC joint in providing stability to the joint (figure 3.24).

Posterior Muscles of the Shoulder Girdle

The posterior muscles of the shoulder girdle include the levator scapulae, rhomboids, and trapezius (figures 3.24).

> **Levator scapulae:** This muscle's name tells the story of its function—to lift the scapula. The levator scapulae muscle originates on the transverse processes of the first four cervical vertebrae

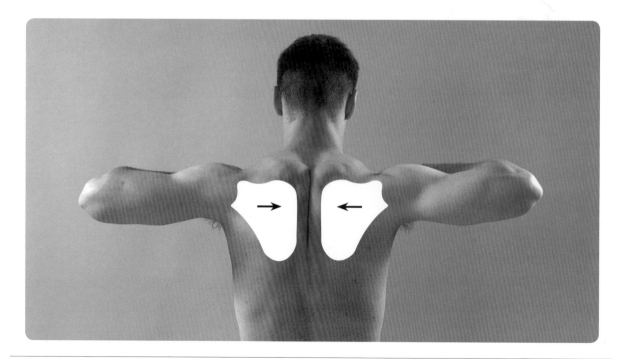

Figure 3.23   Adduction of the scapulas.

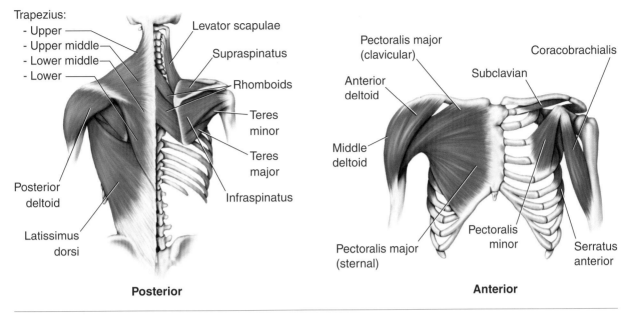

Posterior   **Anterior**

Figure 3.24 The muscles acting at the scapula and humerus.

and inserts on the superior aspect of the vertebral border of the scapula. Because the cervical attachments are the more stable end of this muscle, contraction of the levator scapulae results in eleva-

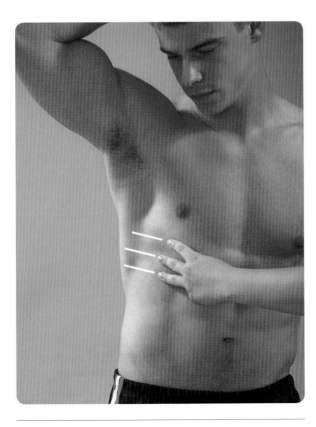

Figure 3.25 Locating the serratus anterior.

tion (downward rotation and adduction) of the shoulder girdle.

> **Rhomboids:** These are actually two muscles (major and minor) that are usually considered as one because they both perform the same function. They originate on the spinous processes of the seventh cervical through fifth **thoracic vertebrae** and insert on the vertebral (medial) border of the scapula. Because the spinous process attachments are the most stable, contraction of the rhomboids results in elevation and adduction (downward rotation) of the shoulder girdle.

> **Trapezius:** This large, triangular muscle originates on the external occipital protuberance at the base of the skull and the spinous processes of all cervical and thoracic vertebrae and inserts on the spine of the scapula and posterior surface of the clavicle. Because of the size of the muscle, the angle of its various fibers, and its numerous functions, discussion of the muscle's action usually divides the muscle into four separate sections (upper, upper middle, lower middle, and lower). The upper fibers of the trapezius parallel to a great degree the levator scapulae muscle and therefore perform similar functions: elevation and adduction (downward rotation) of the shoulder girdle. The upper middle fibers of the trapezius also assist in elevation of the shoulder girdle but function to a greater degree in adduction of the shoulder girdle.

The lower middle fibers of the trapezius function almost exclusively in adduction of the shoulder girdle. The lower fibers of the trapezius, because of the angle at which they run from the thoracic spine to the spine of the scapula, contribute to abduction (upward rotation) of the shoulder girdle.

Hands on . . . On your partner, locate the four portions of the trapezius muscle (figure 3.26).

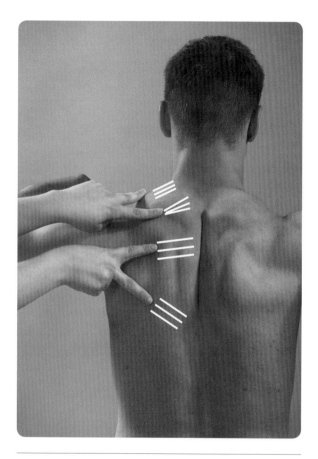

Figure 3.26   Locating the four parts of the trapezius.

Fundamental Movements and Muscles of the Shoulder Joint

We first observe the fundamental movements of the shoulder joint and then look at the muscles responsible for those movements.

Movements of the Shoulder Joint

The shoulder (glenohumeral) joint (figure 3.27) is classified as a triaxial joint because it is capable of movement in all three cardinal planes: flexion (anterior movement of the arm) and extension (return from flexion) in the sagittal plane about a frontal horizontal axis, abduction (movement away from the midline of the body) and adduction (movement toward the midline of the body) in the frontal plane about a sagittal horizontal axis, and internal (inward, medial) and external (outward, lateral) rotation in the horizontal plane about a vertical axis. Because the shoulder joint is a triaxial joint, it is capable of combining fundamental movements to produce **circumduction**. Eleven major muscles function to accomplish the six fundamental movements of the shoulder joint: four anterior, two superior, two posterior, and three inferior to the joint.

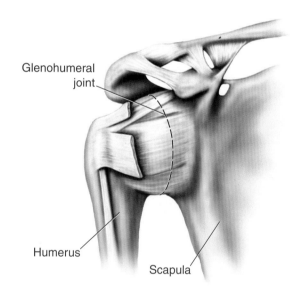

Figure 3.27   Anterior view of the glenohumeral joint.

Anterior Muscles of the Shoulder Joint

The following muscles appear on the anterior aspect of the shoulder joint.

> **Pectoralis major:** The pectoralis major originates on the second to sixth ribs, the sternum, and the medial half of the clavicle and inserts on the anterior area of the surgical neck of the humerus just distal to the greater tuberosity (see figure 3.24). The upper portion of the muscle is often referred to as the clavicular part, and the lower portion is referred to as the sternal part. Contraction of the

pectoralis major muscle produces flexion, adduction, and internal rotation of the shoulder joint.

Hands on . . . Locate your partner's pectoralis major using figure 3.28 as a guide.

Figure 3.28   Locating the pectoralis major.

> **Coracobrachialis:** The coracobrachialis originates on the coracoid process of the scapula (where the tendon of origin is conjoined with the tendon of origin of the short head of the biceps brachii) and inserts on the middle of the medial side of the humerus opposite the deltoid tubercle on the lateral side (figures 3.24 and 3.29). The coracobrachialis flexes the shoulder joint and, because of its angle of pull, assists with adduction of the joint.

Hands on . . . Have your partner place his or her arm at a right angle to the body (abduction) as you palpate the coracobrachialis (figure 3.30).

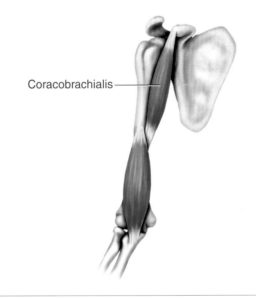

Coracobrachialis —

Figure 3.29   Anterior view of the coracobrachialis.

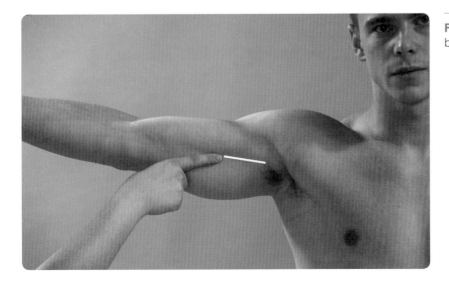

Figure 3.30   Locating the coracobrachialis.

> **Biceps brachii:** Although the biceps brachii is frequently considered a flexor of the elbow (figure 3.31), both the long-head tendon and short-head tendon of the biceps brachii cross the shoulder joint. The long head originates on the supraglenoid tubercle on the superior edge of the glenoid of the scapula, and the short head originates on the coracoid process of the scapula (and is conjoined with the coracobrachialis tendon of origin). Both heads combine to form the belly of the muscle and insert on the tuberosity of the **radius,** which is one of the two forearm bones. Actions produced by contraction of this muscle at the shoulder joint include flexion and abduction by the long-head tendon and flexion, adduction, and internal rotation by the short-head tendon.

> **Subscapularis:** The subscapularis muscle (figure 3.31) is located on the anterior surface of the scapula between the scapula and the thorax. It originates on the large subscapular fossa on the anterior surface of the scapula and inserts on the lesser tuberosity of the humerus. When it contracts, the subscapularis produces internal rotation and flexion at the shoulder joint. This muscle is one of four shoulder joint muscles that attach to a musculotendinous structure often referred to as the **rotator cuff** (discussed later in this chapter).

Superior Muscles of the Shoulder Joint

The following muscles appear on the superior aspect of the shoulder joint.

> **Deltoid:** The deltoid is a very large muscle consisting of three parts: anterior, middle, and posterior (see figure 3.24). It covers the shoulder joint, so it is often referred to as the shoulder cap muscle. The anterior (clavicular) fibers originate from the lateral portion of the anterior aspect of the clavicle, the middle (acromial) fibers originate from the lateral edge of the acromion process of the scapula, and the posterior (scapular) fibers originate on the inferior edge of the spine of the scapula. All three portions combine to insert on the deltoid tubercle on the lateral surface of the middle of the humerus. Contraction of the entire deltoid muscle results in abduction of the shoulder joint; contraction of the posterior portion alone results in adduction, extension, and external rotation; and contraction of the anterior fibers alone results in adduction, flexion, and internal rotation. The middle fibers of the deltoid muscle are typically considered to be involved only in shoulder joint abduction. Once the arm has been abducted to the horizontal level from the anatomical position, all three portions of the muscle are considered abductors of the joint.

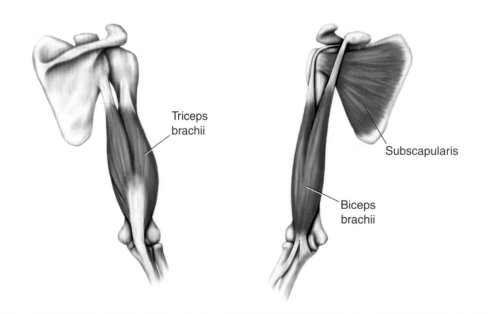

Triceps brachii

Subscapularis

Biceps brachii

Figure 3.31   Anterior and posterior views of two superficial shoulder joint muscles and the subscapularis.

Hands on . . . Have your partner abduct his or her arm, preferably with some form of resistance, such as holding a book or a weight, as you locate all three portions of the deltoid muscle (figure 3.32).

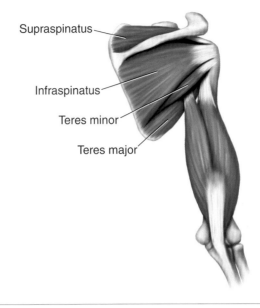

Figure 3.33 Posterior view of the supraspinatus, infraspinatus, teres minor, and teres major muscles.

Posterior Muscles of the Shoulder Joint

The following muscles are found on the posterior aspect of the shoulder joint.

> **Infraspinatus:** The infraspinatus muscle (figure 3.33) gets its name from the anatomical structure where it originates: the infraspinous fossa beneath the inferior surface of the spine of the scapula. The infraspinatus muscle inserts on the middle facet of the greater tuberosity of the humerus. Contraction of the infraspinatus muscle produces external rotation and extension of the shoulder joint. The infraspinatus muscle is also part of the rotator cuff.

Hands on . . . Place your partner's shoulder joint in abduction, external rotation, and extension and then locate the infraspinatus muscle (figure 3.34).

> **Teres minor:** The teres minor muscle is often considered together with the infraspinatus muscle because they share the same function. The teres minor (figure 3.33) originates on the upper and middle portions of the lateral border of the scapula and inserts on the distal facet of the greater tuberosity of the humerus. Contraction of the teres minor muscle, like the infraspinatus muscle, produces external rotation and extension

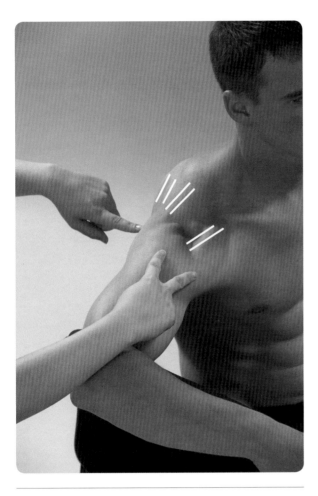

Figure 3.32 Locating the deltoid.

> **Supraspinatus:** Located beneath the deltoid muscle, the supraspinatus muscle (figure 3.33) originates on the supraspinous fossa of the scapula and inserts on the proximal facet of the greater tuberosity of the humerus. The muscle abducts the shoulder joint. Although it contracts throughout the full range of abduction, it is considered the primary initiator of abduction until approximately 30° of abduction, when the deltoid muscle takes over as the major abductor. The supraspinatus muscle is also one of the muscles of the rotator cuff.

joint is so shallow and thus provides little stability. In an action such as throwing, not only do the muscles of the rotator cuff generate the force necessary to throw, but they also decelerate the force generated. In other words, the action of the rotator cuff muscles actually prevents the entire upper extremity from following the object thrown by keeping the humeral head in the glenoid fossa.

Figure 3.34   Locating the infraspinatus and teres minor.

of the shoulder joint. This muscle is also one of the shoulder joint muscles of the rotator cuff.

Hands on . . . Place your partner's shoulder joint in abduction, external rotation, and extension and then locate the teres minor muscle (figure 3.34).

Rotator Cuff

Four of the muscles of the shoulder joint insert on a musculotendinous structure running between the facets located on the lesser and greater tuberosities of the humerus. This structure is commonly referred to as the rotator cuff (figure 3.35). The motions produced at the shoulder joint by these four muscles (subscapularis, supraspinatus, infraspinatus, and teres minor) have been presented, but these four muscles are also responsible for maintaining the stability of the shoulder joint, which is particularly necessary because the socket of this ball-and-socket

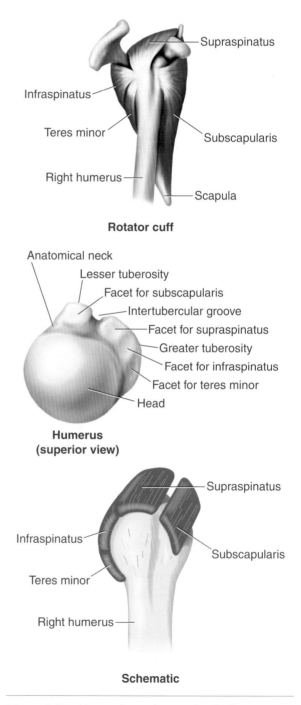

Figure 3.35   Humeral attachments for the four muscles of the rotator cuff.

Focus on . . . THE ROTATOR CUFF

Understanding the action of the rotator cuff makes it easier to understand why some individuals involved in repetitive throwing activities (e.g., pitchers, quarterbacks) develop problems of the rotator cuff. The muscles not only generate the force needed by contracting concentrically (shortening) but also apply a braking action through an eccentric (lengthening) contraction to prevent the upper extremity from leaving the body.

Inferior Muscles

The following muscles cross the shoulder joint inferiorly (underneath).

> **Latissimus dorsi:** The latissimus dorsi muscle (see figure 3.24), a large muscle of the back, originates on the spinous processes of the lower six thoracic and all five **lumbar vertebrae,** the posterior aspect of the **ilium** (see chapter 7), the lower three ribs, and the inferior angle of the scapula; passes beneath the axilla (armpit); and inserts on the edge of the intertubercular groove on the anterior aspect of the humerus. Contraction of the latissimus dorsi muscle produces internal rotation, extension, and adduction of the shoulder joint.

Hands on . . . Place your partner's arm in the position of external rotation and abduction and locate the latissimus dorsi muscle (figure 3.36).

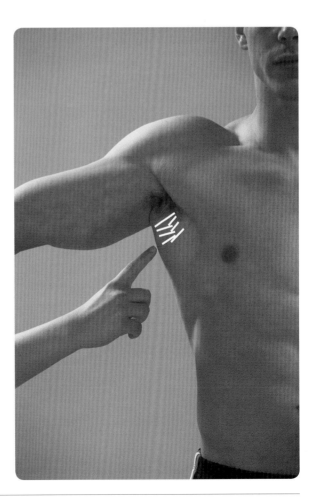

Figure 3.36 Locating the latissimus dorsi.

> **Teres major:** The teres major muscle (figure 3.33) originates on the lower portion of the lateral border of the scapula and its inferior angle, crosses beneath the axilla (armpit), and inserts on the area just inferior to the lesser tuberosity of the humerus. Contraction of the teres major produces the same action as the latissimus dorsi: internal rotation, extension, and adduction of the shoulder joint. Because the action of these two muscles is identical, the teres major is often called the latissimus dorsi's "little helper."

Hands on . . . Placing your partner's arm in the position of abduction, observe the posterior aspect of the shoulder joint and locate the teres major muscle (figure 3.37).

> **Triceps brachii:** Although the triceps brachii is more often associated with elbow joint action, one of the three tendinous heads of the triceps brachii does cross the shoulder joint and assist with shoulder joint movement (see figure 3.31). Of the lateral-head, long-head, and medial-head tendons of origin of the triceps brachii, the long head originates on the infraglenoid tubercle of the glenoid lip of the scapula and joins the lateral and medial heads to insert, on a common tendon, on the olecranon process of the **ulna,** one of the two bones of the forearm. Contraction of the long head of the triceps brachii assists with shoulder joint extension and adduction.

Hands on . . . On yourself or a partner, locate all three portions of the triceps brachii muscle (figure 3.38).

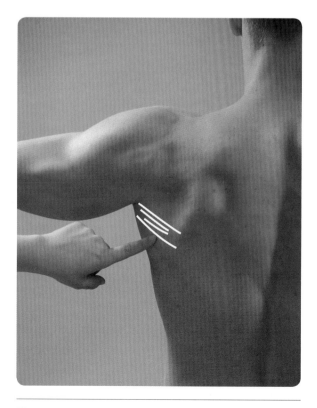

Figure 3.37 Locating the teres major.

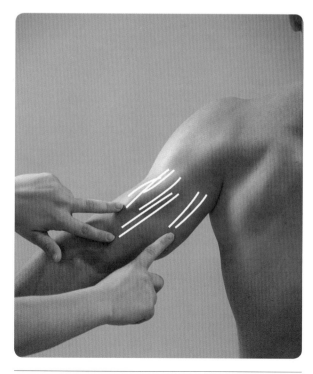

Figure 3.38 Locating the three sections of the triceps brachii.

FOCUS ON . . . THE GLENOHUMERAL JOINT

A dislocation of any joint usually results in the severe spraining of the ligaments and straining of the muscles that cross the joint. Following are three of the more common dislocations of the glenohumeral joint.

Excessive movement in any joint may result in stress being placed on the ligamentous structures tying the bones of the joint together. As noted earlier regarding the AC joint of the shoulder girdle, a sprain of a ligament is the partial or complete tearing of a ligament. If the ligaments of a joint are disrupted to the extent that the bones of the joint actually displace, this is known as a dislocation. Dislocations of the glenohumeral joint are not uncommon. The three most common forms of glenohumeral dislocations are anterior (subcoracoid), posterior (subspinous), and downward (subglenoid) dislocation (see figure 3.39). Anterior dislocation is the most common form of shoulder dislocation, typically resulting from excessive abduction and external rotation of the shoulder joint. The humeral head displaces anteriorly from the glenoid of the scapula and rests anterior to the glenoid just beneath the coracoid process of the scapula (thus the name subcoracoid dislocation). A second form of shoulder joint dislocation is posterior (subspinous) dislocation, which can result from excessive internal rotation and adduction of the shoulder joint. The humeral head displaces posteriorly from the glenoid of the scapula and rests posterior to the glenoid just beneath the spine of the scapula (thus the name subspinous dislocation). The third form of shoulder joint dislocation is known as a downward (subglenoid) dislocation, which may result from excessive shoulder joint abduction with the humerus abutting against the acromion process of the scapula and the head of the humerus being forced downward (subglenoid) beneath the lower edge of the glenoid of the scapula.

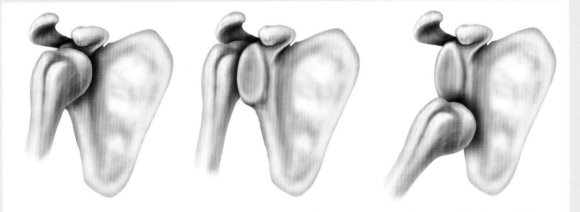

Anterior (subcoracoid)   **Posterior (subspinous)**   **Downward (subglenoid)**

Figure 3.39   Anterior (subcoracoid) glenohumeral dislocation, posterior (subspinous) dislocation, and downward (subglenoid) dislocation.

Combined Actions of the Shoulder Girdle and Shoulder Joint

Each of the shoulder joint actions (flexion, extension, abduction, adduction, internal and external rotation) possesses certain degrees of movement.

When any particular movement of the shoulder joint reaches its end point, to move further in that direction, the glenoid of the scapula must change its position to accommodate additional movement of the humerus. Movement of the shoulder girdle facilitates a greater range of motion in all the fundamental movements of the shoulder joint by changing the position of the glenoid, which is accomplished by movement of the scapula on the thorax (ribs) through motion at the AC and SC joints. The prime example of this relationship is referred to as the **scapulohumeral rhythm.** Although initial abduction of the shoulder joint is attributed to GH joint action only (approximately 120° of abduction), the combination of GH joint abduction and rotation of the scapula results in approximately 180° of shoulder joint abduction. The generally accepted ratio of motion between the GH joint and the scapula (scapulothoracic movement) is that for every 2° of GH joint abduction, the scapula rotates 1°.

As it does with abduction of the shoulder joint, the scapula moves to position the glenoid to accommodate movement to all the other fundamental movements of the shoulder joint. Observe this cooperation between shoulder girdle and shoulder joint movements with the suggested learning activities at the end of the chapter.

REVIEW OF TERMINOLOGY

The following terms were discussed in this chapter. Define or describe each term, and where appropriate, identify the location of the named structure either on your body or in an appropriate illustration.

abduction
acromial end
acromioclavicular (AC) joint
acromioclavicular ligament
acromion process
adduction
anatomical neck
anterior fibers (of the sternoclavicular ligament)
articular disc
biceps brachii
capsular ligament
circumduction
clavicle
clavicular notch
conoid ligament
conoid tubercle
coracoacromial ligament
coracobrachialis
coracoclavicular joint
coracoclavicular ligament
coracohumeral ligament
coracoid process
costal tuberosity
costoclavicular ligament
deltoid
deltoid tubercle (of the clavicle)
deltoid tuberosity (of the humerus)
depression
dislocation
distal facet
downward rotation
elevation
glenohumeral (GH) joint
glenohumeral ligament
glenoid fossa

glenoid lip (glenoid labrum)
greater tuberosity
great scapular notch
head (of the humerus)
humerus
ilium
inferior angle
inferior glenohumeral ligament
inferior transverse scapular ligament
infraglenoid tubercle
infraspinatus
infraspinous fossa
interclavicular ligament
intertubercular groove
lateral border
lateral tilt
latissimus dorsi
lesser tuberosity
levator scapulae
lumbar vertebra
medial border
medial tilt
middle facet
middle glenohumeral ligament
pectoralis major
pectoralis minor
posterior fibers (of the sternoclavicular ligament)
protraction
proximal facet
radius
retraction
rhomboids
rib
rotator cuff
scapula

scapular notch
scapulohumeral rhythm
serratus anterior
shoulder girdle
shoulder joint
spine (of the scapula)
sternal end
sternoclavicular (SC) joint
sternoclavicular ligament
sternum
subclavian groove
subclavian muscle
subscapular fossa
subscapularis
superior angle
superior border
superior fibers (of the sternoclavicular ligament)
superior glenohumeral ligament
superior transverse scapular ligament
supraglenoid tubercle
supraspinatus
supraspinous fossa
surgical neck
teres major
teres minor
thoracic vertebra
thorax
transverse humeral ligament
trapezius
trapezoid ligament
trapezoid line
triceps brachii
ulna
upward rotation

SUGGESTED LEARNING ACTIVITIES

1. Place your hands on a partner's scapula. Ask the partner to slowly abduct both shoulder joints. As the humerus moves away from the body, determine when the scapula starts to move. Did the scapula move throughout abduction of the shoulder joint? When did it start to move? Why did it move? What muscle initiated this action? Repeat this activity during shoulder joint flexion, hyperextension, and internal and external rotation and ask yourself these same questions.

2. Ask a partner in anatomical position to abduct both shoulder joints to the point where his or her hands touch, while you place your hands on both scapulas. What movement did the scapulas perform? Did the partner reach the end point through just abduction, or did the partner have to rotate each shoulder joint? If so, why was this necessary? In what direction?

3. Flex your partner's elbow to 90°. Passively abduct and externally rotate the shoulder joint to the point where you feel resistance. What muscles are providing that resistance? (Do this activity gently. Excessive force can harm the joint structures.)

4. Flex your partner's elbow to 90°. Passively adduct and internally rotate the shoulder joint to the point where you feel resistance. What muscles are providing that resistance? (Do this activity gently. Excessive force can harm the joint structures.)

5. Place your hands on top of your partner's shoulders (hands on clavicle and scapula) and push downward while asking your partner to elevate (shrug) the shoulders. What muscles performed this activity? What muscles are being used to oppose your resistance?

MULTIPLE-CHOICE QUESTIONS

1. In the anatomical position, depression of the shoulder girdle is defined as
   a. active
   b. passive
   c. isometric
   d. impossible

2. The only portion of the triceps brachii muscle that crosses the shoulder joint is the
   a. medial-head tendon
   b. lateral-head tendon
   c. short-head tendon
   d. long-head tendon

3. Which part of the deltoid muscle is involved only in shoulder joint abduction?
   a. middle fibers
   b. posterior fibers
   c. anterior fibers
   d. inferior fibers

4. In addition to elevation of the scapula, the levator scapulae muscle also performs what other shoulder girdle action?
   a. upward rotation
   b. flexion
   c. downward rotation
   d. extension

5. The articulation formed by the clavicle and the scapula is often referred to as the
   a. AC joint
   b. SC joint
   c. GH joint
   d. SH joint

6. Movement of the scapula away from the midline of the body is defined as
   a. upward rotation
   b. adduction
   c. downward rotation
   d. elevation

7. Which of the following muscles does not have a role in the rotator cuff, which provides both motion and stability at the glenohumeral joint?
   a. supraspinatus
   b. subscapularis
   c. infraspinatus
   d. teres major

8. The glenohumeral joint is which type of joint?
   a. nonaxial
   b. uniaxial
   c. biaxial
   d. triaxial

(continued)

9. What muscle, known as the latissimus dorsi's "little helper," extends, adducts, and internally rotates the shoulder joint?

   a. supraspinatus
   b. infraspinatus
   c. teres minor
   d. teres major

10. The primary movement of the shoulder girdle produced by the contraction of the rhomboids is

    a. abduction
    b. adduction
    c. upward rotation
    d. depression

11. Which of the following muscles is not considered an anterior muscle of the shoulder joint?

    a. pectoralis major
    b. pectoralis minor
    c. subscapularis
    d. coracobrachialis

12. The short head of the biceps brachii originates on the

    a. acromion process
    b. glenoid fossa
    c. coracoid process
    d. greater tuberosity

13. When the humerus is abducted to the point that the arm is held upright over one's head, the scapula is

    a. abducted
    b. adducted
    c. flexed
    d. extended

14. The inferior angle, the medial border, the lateral border, and the spine are all bony landmarks associated with which of the following bones?

    a. sternum
    b. clavicle
    c. scapula
    d. humerus

15. Which of the following ligaments is not present at the AC joint of the shoulder girdle?

    a. acromioclavicular
    b. trapezoid
    c. conoid
    d. costoclavicular

16. The coracoid process of the scapula serves as the attachment for the conjoined tendon of the coracobrachialis muscle and what other muscle?

    a. long head of the biceps
    b. brachioradialis
    c. brachialis
    d. short head of the biceps

17. Which of the following muscles is not considered a posterior muscle of the shoulder girdle?

    a. rhomboids
    b. latissimus dorsi
    c. trapezius
    d. levator scapulae

18. The primary function of the biceps brachii at the shoulder joint is flexion, but the long-head tendon of the biceps brachii also assists with what other movement of the shoulder joint?

    a. abduction
    b. adduction
    c. extension
    d. external rotation

19. The serratus anterior muscle is primarily involved in what shoulder girdle action?

    a. abduction
    b. adduction
    c. elevation
    d. depression

20. The infraspinatus muscle and which of the following muscles are usually considered the primary external rotators of the shoulder joint?

    a. supraspinatus
    b. subscapularis
    c. teres major
    d. teres minor

21. The rhomboid muscles help elevate the scapula as well as rotate it downward to produce what other scapular movement?

    a. abduction
    b. adduction
    c. flexion
    d. extension

22. The lesser tuberosity of the humerus serves as the source of attachment for which of the following muscles?

    a. infraspinatus
    b. subscapularis
    c. supraspinatus
    d. teres minor

23. Which of the following shoulder girdle actions is performed by the pectoralis major?

    a. upward rotation
    b. adduction
    c. elevation
    d. none

24. The lowest part of the trapezius muscle assists with which of the following movements of the shoulder girdle (when performed from the anatomical position)?
    a. elevation
    b. depression
    c. downward rotation
    d. upward rotation

25. For movement of the shoulder joint to occur in the frontal plane, which of the following joint actions must take place?
    a. internal rotation
    b. circumduction
    c. flexion
    d. abduction

26. The costoclavicular ligament is found in which of the following joints of the shoulder girdle?
    a. sternoclavicular
    b. humeroclavicular
    c. acromioclavicular
    d. thoracoclavicular

27. One of the tendons of which of the following muscles lies in the anatomical structure known as the intertubercular groove?
    a. biceps brachii
    b. coracobrachialis
    c. deltoid
    d. supraspinatus

FILL-IN-THE-BLANK QUESTIONS

1. Two ligaments known as the conoid and the trapezoid ligaments, which join the scapula and the clavicle together, are commonly called the _____ ligament.

2. The muscle that originates from the base of the skull (external occipital protuberance) to the end of the thoracic vertebrae (approximately two thirds of the way down the back) is known as the _____ _____.

3. A muscle of the rotator cuff that crosses the anterior portion of the shoulder joint and is a major internal rotator of the shoulder joint is the _____ _____ muscle.

4. The muscle known as the initiator of shoulder joint abduction is the _____ muscle.

5. At the lateral end of the spine of the scapula, there is a wide, bony projection of the spine known as the _____.

6. The muscle that runs between the coracoid process of the scapula and the medial surface of the humerus opposite the deltoid tuberosity is known as the _____ muscle.

7. The glenoid fossa (shoulder joint socket) is located on the lateral aspect of the _____.

8. When all three portions of the deltoid muscle contract together, the shoulder joint moves into _____.

9. The pectoralis major muscle has two distinct parts: a lower portion known as the sternal part and a superior portion known as the _____ part.

10. The coracoid process is found on the _____ _____.

11. A broad superficial muscle of the lower back lateral and inferior to the trapezius that is a powerful internal rotator of the shoulder joint is known as the _____ muscle.

CHAPTER FOUR

The Elbow and Forearm

The elbow and the forearm are composed of three bones: the **humerus,** the **ulna,** and the **radius.** Together these three bones form four joints, three at the proximal end of the forearm (**radiohumeral, ulnohumeral,** and **proximal radioulnar**) and one at the distal end of the forearm (**distal radioulnar**).

Bones of the Elbow and Forearm

The humerus, the ulna, and the radius meet to form the structure commonly known as the **elbow,** and the ulna and radius form the **forearm** (figure 4.1). The proximal end of the humerus was discussed in chapter 3 on the shoulder joint. The distal end of the humerus provides the bony attachments for the soft tissues that span the upper arm and the forearm to form the elbow joint. At the distal end of the humerus, the shaft widens out to form two bony prominences: the **lateral** and **medial epicondyles.**

Hands on . . . Find these structures on your own elbow or your partner's elbow (figure 4.2). These epicondyles are easily palpated and are visible on most individuals.

The parts of the humerus between the shaft and these epicondyles, where the bone actually widens, are known as the **lateral** and **medial supracondylar ridges.** There are three fossa (depressions) on the distal humerus that provide areas for other bony structures to move into during elbow joint actions. On the anterior surface, between the two epicondyles, is the **coronoid fossa.** This is where the coronoid process of the ulna is positioned during elbow flexion. Lateral to the coronoid fossa is the **radial fossa.** The head of the radius moves into this fossa during elbow flexion. Just distal to the lateral epicondyle is a smooth, round surface known as the **capitulum.** This surface is where the head of the radius rotates during forearm movement. At the very distal end of the humerus is a spool-like structure known as the **trochlea.** This is the

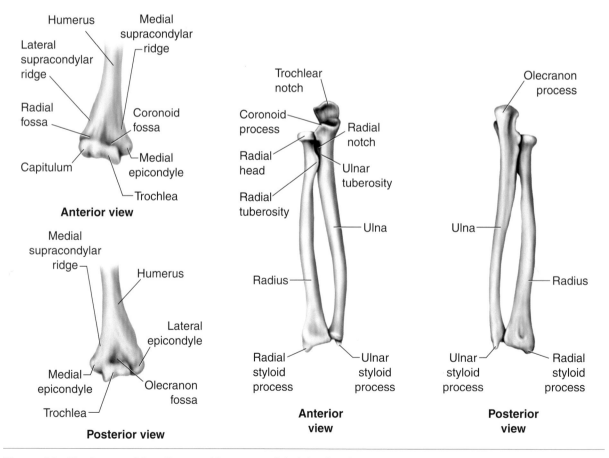

Figure 4.1   The bones of the elbow and forearm and their landmarks.

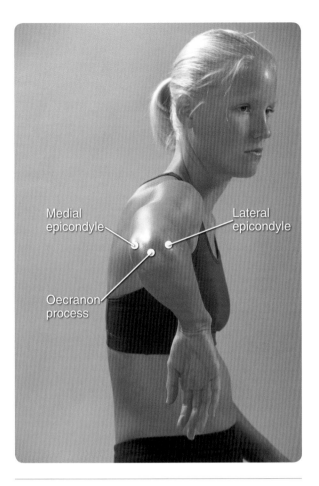

Figure 4.2 Locating the medial and lateral epicondyles.

structure on which the olecranon process of the ulna attaches.

A posterior view of the distal end of the humerus also reveals some structures previously discussed: the supracondylar ridges, the epicondyles, and the trochlea. In addition, the **olecranon fossa,** which is formed between the two epicondyles, appears. This is the depression into which the **olecranon process** of the ulna moves when the elbow joint is moved into extension.

Under normal conditions, when the elbow joint is in full extension, you should be able to observe that the lateral epicondyle, the olecranon process, and the medial epicondyle form a straight line. If the normal elbow joint is flexed to 90°, these three structures should form an isosceles triangle with the olecranon process distal to the epicondyles (figure 4.2).

The two bones of the forearm are known as the radius and the ulna. From the anatomical posi-

tion, the radius is on the lateral aspect and the ulna on the medial aspect of the forearm. In the elbow joint, the ulna has the prominent role of articulating with the humerus, whereas the radius plays the more prominent role of articulating with bones of the wrist. The proximal end of the radius consists of the **radial head,** the **radial neck,** and a large **radial tuberosity** distal to the neck on the medial aspect of the upper portion of the shaft of the bone.

Hands on . . . You can easily palpate the head of the radius at the lateral aspect of your elbow (figure 4.3).

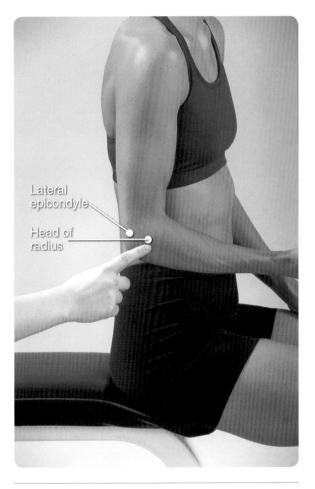

Figure 4.3 Locating the radial head.

At the distal end of the shaft of the radius, on the lateral aspect, is a large prominence known as the **radial styloid process** (also known as the lateral styloid process).

Hands on . . . This structure can be easily palpated in the area where the hand and forearm come together (wrist) on the thumb side (figure 4.4).

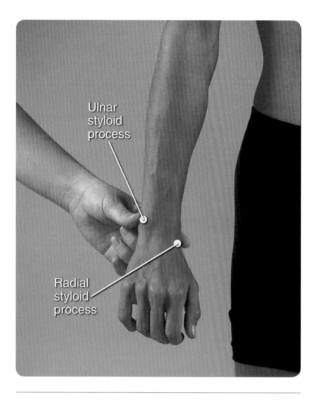

Figure 4.4 Locating the ulnar and radial styloid processes.

On the medial aspect of the distal end of the radius is the **ulnar notch.** On the distal surface of the radius there are two distinct facets where bones of the wrist articulate (figure 4.5).

The medial forearm bone, the ulna, has a very large prominence at the proximal end known as the olecranon process (see figure 4.1). This structure has a cuplike surface known as the trochlear notch that rotates about the trochlea of the humerus to form the articulation between the humerus and the ulna commonly known as the elbow joint. At the very anterior portion of the trochlear notch is a smaller prominence known as the **coronoid process.** Just lateral to the coronoid process is the **radial notch** of the ulna. The head of the radius articulates with the ulna (proximal radioulnar joint) at the radial notch of the ulna.

At the distal end of the ulna on the medial aspect is an easily palpated prominence known as the **ulnar styloid process** (also called the medial styloid process; see figures 4.1 and 4.4).

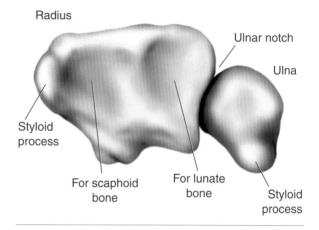

Figure 4.5 Facets on the distal surface of the radius where bones of the wrist articulate.

Ligaments and Joints of the Elbow and Forearm

With three bones (humerus, ulna, and radius) coming together to form the elbow joint (figure 4.6), there are actually three joints in the anatomical area between the upper arm and the forearm: the "true" elbow joint between the humerus and the ulna (the ulnohumeral joint), the radiohumeral joint between radius and humerus, and the proximal radioulnar joint between radius and ulna. The **capsular ligament** surrounds all three of these articulations. It is divided into an anterior and posterior part. The anterior part extends from the anterior surface of the humerus just proximal to the coronoid fossa to the anterior surface of the coronoid process and the **annular ligament.** Laterally, the capsular ligament fuses with the **collateral ligaments.** Posteriorly, the capsular ligament attaches to the tendon of insertion of the triceps brachii muscle, the edge of the olecranon, the lateral epicondyle, and posterior surface of the humerus in the area of the trochlea and the capitulum. Distally, the capsular ligament attaches to the lateral and superior edges of the olecranon process, the posterior aspect of the annular ligament, and posterior to the radial notch of the ulna.

The ligaments that fuse with the anterior portion of the capsular ligament are known as the **radial** (lateral) **collateral ligament** and the **ulnar** (medial) **collateral ligament.** The radial collateral ligament runs between the inferior border of the lateral epicondyle of the humerus to the annular ligament and the radial notch of the ulna.

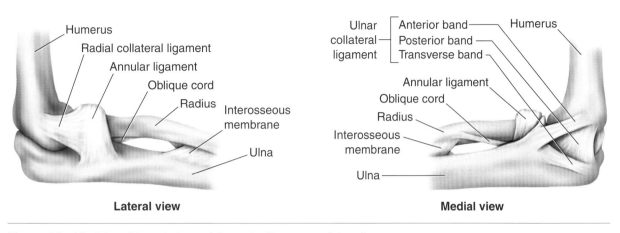

Lateral view

Medial view

Figure 4.6 Medial and lateral views of the major ligaments of the elbow.

The ulnar collateral ligament has three distinct parts (see figure 4.6): an **anterior band** that runs between the anterior inferior area of the medial epicondyle of the humerus and the medial aspect of the coronoid process of the ulna, a **posterior band** running between the medial epicondyle of the humerus and the medial border of the olecranon process of the ulna, and the **transverse band,** which does not cross the elbow joint, running between the anterior band on the coronoid process of the ulna and the posterior band on the olecranon process of the ulna.

The remaining joint of the elbow, the proximal radioulnar joint, is actually a joint between the bones of the forearm. The joint is between the head of the radius and the radial notch of the ulna. The annular ligament runs between the anterior and posterior edge of the radial notch on the ulna and forms a ring completely around the head of the radius.

Running between the shafts of the ulna and radius is a ligamentous band of connective tissue known as the **interosseous membrane** that helps distribute pressure between the ulna and radius when force is applied and also serves as a source of muscular attachment for several forearm muscles. At the proximal end of the interosseous membrane is a ligament known as the **oblique cord** that functions to prevent separation between the ulna and radius.

At the distal end of the ulna and radius, just proximal to the wrist, is the distal radioulnar joint. In the concave ulnar notch of the radius, the round head of the ulna rotates, forming a pivot joint. The ligaments of this joint are the **dorsal radioulnar** and the **volar radioulnar** (figure 4.7).

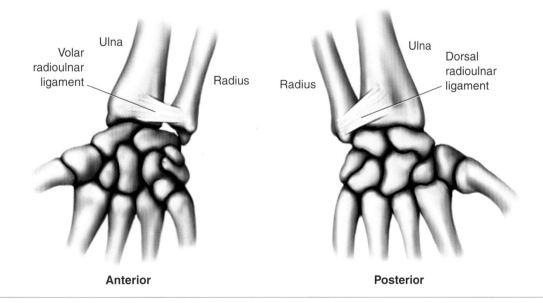

Anterior

Posterior

Figure 4.7 The dorsal and volar radioulnar ligaments.

Fundamental Movements and Muscles of the Elbow and Forearm

Remember that the starting position for description of all fundamental movements of any joint is the anatomical position. The elbow joint (articulation between the trochlea of the humerus and the olecranon process of the ulna) is a uniaxial joint capable of flexion and extension in the sagittal plane about a frontal horizontal axis. Five major muscles produce the motions of flexion and extension of the elbow joint. They are the **brachialis** (flexion), the **brachioradialis** (flexion), the **biceps brachii** (flexion), the **triceps brachii** (extension), and the **anconeus** (extension). The brachialis, brachioradialis, and biceps brachii muscles are anterior to the elbow joint, and the triceps brachii and anconeus muscles are posterior to the joint. Four muscles are responsible for the movements of supination and pronation of the forearm (figure 4.8). The biceps brachii (supination) has already been mentioned in its other role at the elbow joint (flexion). The other three muscles involved with forearm motion are the **supinator** (supination), the **pronator quadratus** (pronation), and the **pronator teres** (pronation).

Anterior Muscles of the Elbow

The anterior muscles of the elbow are three in number: the brachialis, the brachioradialis, and the biceps brachii. Two of these muscles (the brachialis and brachioradialis) are involved exclusively in one movement of the elbow joint, whereas the third (the biceps brachii) is involved in elbow joint movement and also movement of the forearm.

> **Brachialis:** The brachialis muscle originates on the middle of the anterior shaft of the humerus and inserts on the coronoid process of the ulna. Because of its origin and insertion, its only function is flexion of the elbow joint. It is located beneath the biceps brachii muscle (figures 4.9 and 4.10).

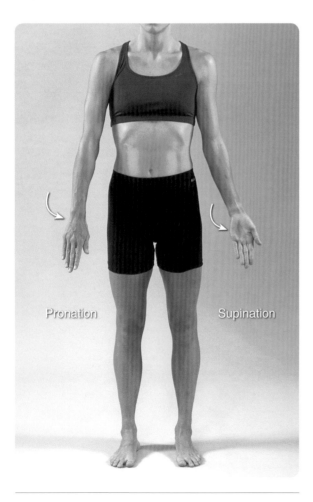

Figure 4.8 Pronation and supination of the radioulnar joints.

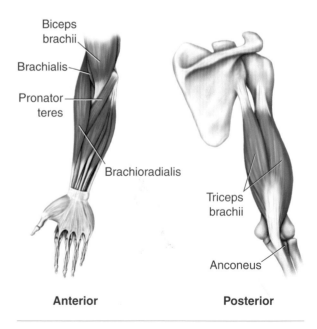

Figure 4.9 Superficial muscles of the elbow and forearm.

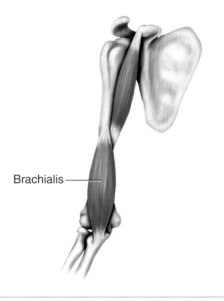

Brachialis

Figure 4.10 Anterior view of the brachialis.

Hands on . . . Have your partner abduct and externally rotate his or her shoulder joint so you can locate the brachialis muscle (figure 4.11).

> **Brachioradialis:** The brachioradialis muscle originates on the lateral epicondyle of the humerus and inserts on the radial styloid process. The muscle crosses the anterior aspect of the elbow joint and therefore is a flexor of the elbow joint (figure 4.9).

Hands on . . . Have your partner flex his or her elbow joint against a slight resistance applied by the opposite hand as you locate the brachioradialis muscle (figure 4.12).

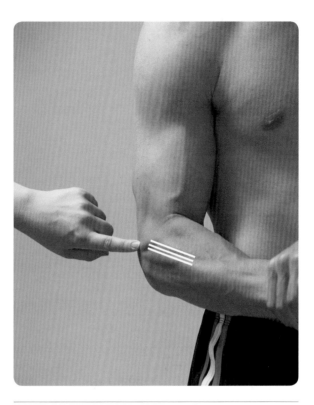

Figure 4.12 Locating the brachioradialis.

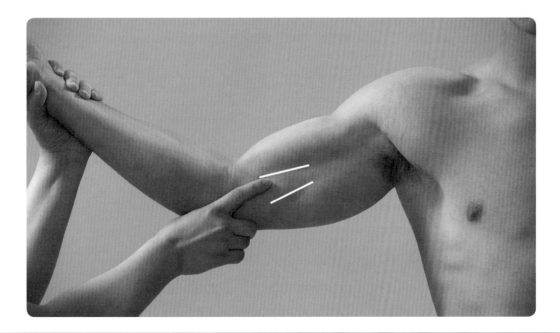

Figure 4.11 Identifying the brachialis.

Figure 4.13 Demonstrating the biceps brachii.

> **Biceps brachii:** The biceps brachii is commonly considered a flexor of the elbow (see figure 4.9), but both the long-head tendon and short-head tendon of the biceps brachii also cross the shoulder joint and contribute to shoulder motion (figure 4.13). The long head originates on the supraglenoid tubercle on the superior edge of the glenoid of the scapula, and the short head originates on the coracoid process of the scapula (and is conjoined with the coracobrachialis tendon of origin). Both heads combine into the belly of the muscle, which inserts on the tuberosity of the radius. Actions produced by contraction of this muscle are flexion at the elbow joint and supination at the forearm.

Posterior Muscles of the Elbow

The anconeus and triceps brachii are posterior muscles of the elbow.

> **Triceps brachii:** The triceps brachii is most often associated with elbow joint extension (see figure 4.9), but one of its three tendinous heads crosses the shoulder joint and assists with shoulder joint movement. The long head originates on the infraglenoid tubercle of the glenoid lip of the scapula and joins the lateral and medial heads to insert, on a common tendon, on the olecranon process of the ulna. Contraction of the triceps brachii extends the elbow joint.

Hands on . . . To palpate the three heads of the triceps brachii (the lateral head, long head, and medial head), have a partner extend his or her elbow against a resistance while also extending the shoulder joint, and palpate all three heads (figure 4.14).

> **Anconeus:** The anconeus originates on the lateral epicondyle of the humerus and inserts on the olecranon process of the ulna. This muscle assists the triceps brachii in elbow extension (see figure 4.9). A common error is to assume that the anconeus is also involved in movement of the forearm; on careful inspection, one can observe that this muscle's origin and insertion points allow only one action when the muscle contracts: elbow extension.

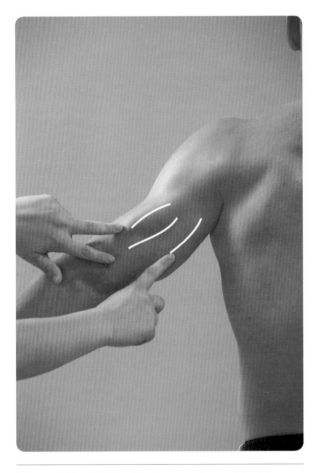

Figure 4.14 Finding the three sections of the triceps brachii.

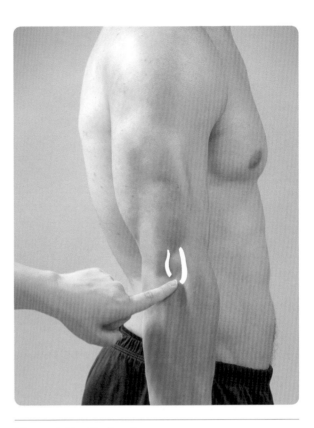

Figure 4.15 Finding the anconeus.

Hands on . . . Locate the anconeus muscle on the posterior lateral aspect of your or your partner's elbow (figure 4.15).

Muscles of the Forearm

Muscles of the forearm include two pronator and two supinator muscles. All are major producers of forearm movement. Only the biceps brachii has another function, that of elbow flexion, as previously mentioned.

> **Pronator teres:** Originating on the coronoid process of the ulna and inserting on the lateral surface of the radius, the pronator teres, as indicated by its name, is responsible for pronation of the forearm (see figure 4.9).

Hands on . . . Manually resist your partner's attempt to flex his or her elbow and pronate his or her wrist while you look for the pronator teres muscle (figure 4.16).

Figure 4.16 Finding the pronator teres muscle.

Focus on . . . THROWING ACTION

The pronator teres muscle and a group of muscles that originate in the common flexor tendon, presented in chapter 5 on the wrist and hand, come under great stress in the overhead throwing motion, particularly at the end of the windup stage of throwing. This pronator–flexor muscle group also is stressed at the release stage of throwing a curve ball in baseball pitching. Inflammation of the pronator–flexor group at its origin in the area of the medial epicondyle of the humerus is often referred to as "pitcher's elbow" or "Little League elbow."

A further complication from this throwing action is the spraining and possible complete disruption of the medial or ulnar collateral ligament of the elbow joint. In 1974, Dr. Frank Jobe, a Los Angeles orthopedic surgeon, reconstructed the torn ulnar collateral ligament of major league baseball pitcher Tommy John. Dr. Jobe used a section of the tendon of insertion of the palmaris longus, an anterior muscle of the forearm, and wove it through drill holes in both the humerus and the ulna, making the tendon a substitute for the disrupted ligament. Today, the sports world refers to this reconstruction technique as the "Tommy John surgery."

An understanding of proper throwing mechanics is essential to preventing elbow and shoulder joint problems. Studies in kinesiology, biomechanics, athletic training, and sports medicine provide that understanding.

> **Pronator quadratus:** This muscle originates on the radius and inserts on the ulna just proximal to the wrist (figure 4.17). Its name reflects its function and its shape. Because the ulna is the stable bone in the distal radioulnar articulation, when the pronator quadratus muscle contracts, the radius is pulled over the ulna, and forearm pronation takes place.

> **Supinator:** The supinator originates on the ulna and inserts on the radius at the proximal ends of the bones on the posterior aspect (figure 4.17). The name of this muscle indicates its function: forearm supination.

> **Biceps brachii:** Because of the position of the biceps brachii attachment to the radial tuberosity, when the forearm is in pronation, contraction of the biceps brachii causes the radius to rotate externally (laterally), causing the forearm to supinate (see figures 3.31 and 4.9).

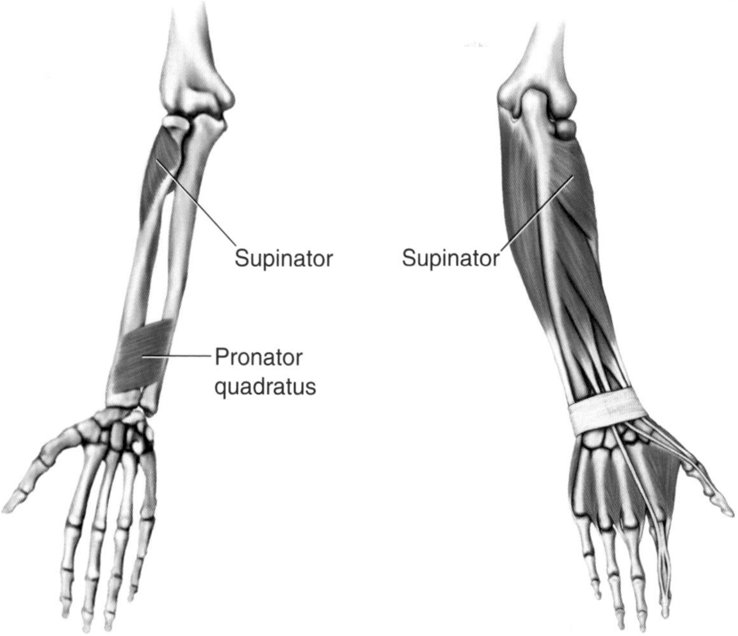

Figure 4.17 The pronator quadratus and the supinator muscles.

Focus on . . . TENNIS ELBOW

"Tennis elbow" is the bane of tennis players. In tennis, when you strike the ball using a backhand stroke, particularly if you attempt to put topspin on the ball as it is struck, the forearm moves from pronation to supination, the wrist moves from flexion to extension, and the supinator muscle along with the muscles that originate in the common extensor tendon (presented in chapter 5 on the wrist and hand) contracts. If the tennis ball is hit off center (outside the so-called sweet spot), a torque (turning force) is applied to the racket that opposes the force applied by the supinator–extensor muscle group. This can result in tennis elbow, an inflammation in the area of the supinator–extensor muscle group's origin on the lateral epicondyle of the humerus. An understanding of proper stroke technique is essential to prevent tennis elbow from developing.

As a challenge, think about how the two-handed backhand stroke might help prevent tennis elbow. Hint: The biomechanical action of the forearm is different in the one-handed and two-handed backstrokes. How is the stress reduced?

REVIEW OF TERMINOLOGY

The following terms were discussed in this chapter. Define or describe each term, and where appropriate, identify the location of the named structure either on your body or in an appropriate illustration.

anconeus
annular ligament
anterior band
biceps brachii
brachialis
brachioradialis
capitulum
capsular ligament
collateral ligaments
coronoid fossa
coronoid process
distal radioulnar joint
dorsal radioulnar ligament
elbow
forearm
humerus

interosseous membrane
lateral epicondyle
lateral supracondylar ridge
medial epicondyle
medial supracondylar ridge
oblique cord
olecranon fossa
olecranon process
posterior band
pronator quadratus
pronator teres
proximal radioulnar joint
radial collateral ligament
radial fossa
radial head
radial neck

radial notch
radial styloid process
radial tuberosity
radiohumeral joint
radius
supinator
transverse band
triceps brachii
trochlea
ulna
ulnar collateral ligament
ulnar notch
ulnar styloid process
ulnohumeral joint
volar radioulnar ligament

SUGGESTED LEARNING ACTIVITIES

1. Either with a skeletal model or a partner, observe the relationship between the lateral and medial epicondyles of the humerus and the olecranon process of the ulna both in the anatomical position and with the elbow flexed to 90°. In which position do these three points form a straight line? In which position do they form an isosceles triangle?

2. Grasp a broom handle, stick, barbell, or similar object with both hands. First with palms up (supination of the forearms) and then with the palms down (pronation of the forearms), perform a biceps curl (flex the elbows from the anatomical position of elbow extension to the position of full elbow flexion). Note the difference in the feeling in your forearms, wrists, and hands when the curl is performed in forearm pronation and forearm supination. What caused the difference when the forearms were in pronation at the beginning of the movement?

(continued)

SUGGESTED LEARNING ACTIVITIES   *(continued)*

3. Apply resistance to your partner's attempt to flex his or her elbow. Place the fingers of your other hand on either side of your partner's biceps brachii muscle. What is the name of the muscle you feel contracting beneath the biceps brachii?

4. Grip a tennis racket or similar object and slowly perform the backstroke. Note the position of your forearm (pronated or supinated) and your wrist (flexed or extended). Stop your stroke at a point where you think the racket will strike the ball. Picture the ball being struck near the upper edge of the racket as opposed to the middle portion of the racket (the so-called sweet spot). Which way will your racket rotate in your grip at the point when the ball is hit? Will your wrist likely be forced to flex or extend on contact with the ball? Will your forearm likely be forced to pronate or supinate on contact? What muscles of the forearm and wrist contract as you perform the backhand stroke? What happens to these muscles if the ball is hit "off center" causing a rotation (torque) of the racket and your forearm? What traumatic condition can be caused by this action if it is repeated every time you strike a backhand stroke? Which of the following factors can affect (positively or negatively) this condition: racket head size, racket grip size, stroke mechanics, a two-handed backstroke?

MULTIPLE-CHOICE QUESTIONS

1. Which of the following anterior muscles is involved in supination of the forearm?

    a. pronator teres
    b. brachioradialis
    c. biceps brachii
    d. brachialis

2. Which of the following muscles assists the triceps brachii in extension of the elbow joint?

    a. extensor digitorum
    b. extensor carpi radialis
    c. anconeus
    d. supinator

3. The relationship between the ulnar (medial) styloid process and the radial (lateral) styloid process is such that the radial (lateral) styloid process is more

    a. proximal
    b. distal
    c. anterior
    d. posterior

4. The ligament running between the shafts of the ulna and radius is known as the

    a. ulnar collateral ligament
    b. radial collateral ligament
    c. annular ligament
    d. interosseous ligament

5. The radial head rotates on what aspect of the humerus?

    a. coronoid fossa
    b. capitulum
    c. olecranon fossa
    d. radial tubercle

6. The structures known as the supracondylar ridges are located where in anatomical relation to the epicondyles?

    a. proximal to the epicondyles
    b. distal to the epicondyles
    c. lateral to the epicondyles
    d. medial to the epicondyles

7. Of the following muscles involved in pronation–supination of the forearm, which crosses the elbow joint?

    a. supinator
    b. biceps brachii
    c. pronator teres
    d. pronator quadratus

8. The lateral collateral ligament of the elbow joint is also known as the

    a. ulnar collateral ligament
    b. radial collateral ligament
    c. annular ligament
    d. interosseous ligament

9. The olecranon process of the elbow joint is located on the posterior aspect of the

    a. distal humerus
    b. proximal humerus
    c. proximal radius
    d. proximal ulna

10. The medial collateral ligament of the elbow joint is also known as the

    a. ulnar collateral ligament
    b. radial collateral ligament
    c. annular ligament
    d. interosseous ligament

11. The pronator–flexor group of forearm muscles originates on the
    a. lateral humeral epicondyle
    b. coronoid process
    c. medial humeral epicondyle
    d. olecranon process

12. The supinator–extensor group of forearm muscles originates on the
    a. lateral humeral epicondyle
    b. coronoid process
    c. medial humeral epicondyle
    d. olecranon process

FILL-IN-THE-BLANK QUESTIONS

1. The olecranon fossa is found on the _____ _____.

2. The ligament that holds the radial head to the ulna is known as the _____ ligament.

3. The coronoid process of the elbow joint is located on the anterior aspect of the _____.

4. The function of the brachialis muscle is _____ _____.

5. The spool-like structure at the distal end of the humerus is known as the _____.

6. When the elbow joint is held in full extension, the olecranon process and the epicondyles form a _____.

7. From the anatomical position, internal rotation of the forearm is called _____.

8. The muscle *chiefly* responsible for the extension of the elbow joint is the _____.

9. A muscle found just proximal to the wrist that is involved in the action that results in the forearm and the palm of the hand being turned downward or backward is the _____.

CHAPTER FIVE

The Wrist and Hand

CHAPTER OUTLINE

The wrist and the hand are complicated structures with multiple bones (figure 5.1), ligaments, joints, and muscles. Because of the fine movements performed by the hand and thumb, these areas are very complex and require more time and effort to learn about than areas previously presented. Although the thumb is often considered one of the hand's five fingers, its movements are unique compared with those of the other four fingers, and it is discussed separately. Humans have **prehensile hands** (i.e., capable of grasping). The capacity to grasp is a direct result of the ability of the thumb to perform opposition, which is discussed under the movements of the thumb. The structure of the thumb joint and its muscles of the thenar eminence (and to a lesser extent, the hypothenar eminence of the little finger) contribute to the ability of primates (the mammalian order that includes monkeys, apes, and human beings) to grasp things in their hands. These structures are presented in detail later in this chapter.

Bones of the Wrist and Hand

The wrist contains eight bones, roughly aligned in two rows, known as the **carpal** bones. The proximal row of carpal bones contains the bones

that articulate with the forearm (radius and ulna), and the distal row of carpal bones articulate with the long bones of the hand (the **metacarpals).** The proximal row of carpal bones, from lateral to medial, are identified as the **scaphoid** (also known as the **navicular),** the **lunate,** the **triquetrum,** and the **pisiform.** The pisiform and the scaphoid bones in the proximal row are easy to palpate.

Hands on . . . Place your index finger on the spots indicated in figures 5.2 and 5.3, and apply pressure downward to feel these bones.

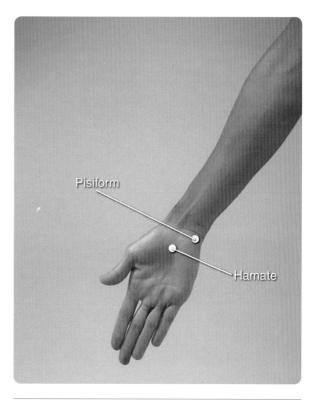

Figure 5.2   Locating the pisiform and the hamate.

The scaphoid (navicular) is a peanut-shaped bone that is the most frequently fractured bone of the wrist when forced into the distal end of the radius, typically from a fall on an extended wrist. The lunate, because of its smooth, dome-shaped proximal end, is the most frequently dislocated bone of the wrist when it is forced into the distal end of the radius, typically from a fall on a flexed wrist. The distal row of carpal bones articulate with the bones of the hand. They are, from lateral to medial, the **trapezium** (also known as the

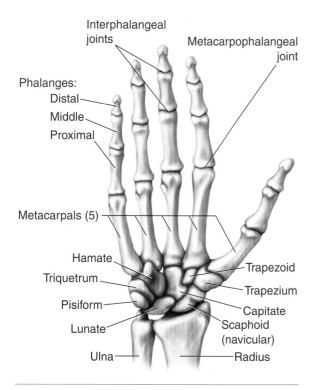

Figure 5.1   Bones of the wrist and hand.

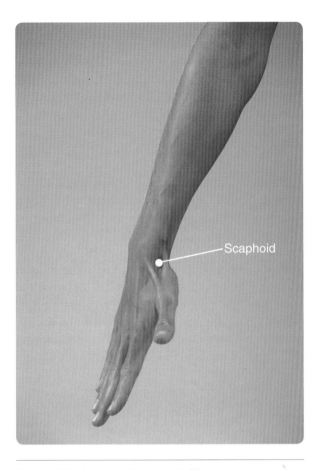

Figure 5.3 Locating the scaphoid.

Joints and Ligaments of the Wrist and Hand

With the large number of bones composing the wrist (ulna, radius, eight carpals, and five metacarpals), it makes sense that there are many, many joints that make up the structure known as the wrist (figure 5.4). There are joints between the forearm bones and the proximal row of the carpals (**radiocarpals**), joints between the proximal and distal rows of carpals (**midcarpals**), and joints between the distal row of carpals and the five metacarpal bones of the hand (**carpometacarpals**). In addition to these wrist joints, there are joints between the carpal bones within each row (**intercarpals**).

What is commonly referred to as the wrist joint is the articulation between the distal end of the radius and primarily two bones of the proximal row of carpal bones: the scaphoid (navicular) and the lunate. The movement between these bones produces a gliding type of action as they roll or slide over each other. The radiocarpal joints are classified as condyloid joints because of their movements.

There are five main ligaments of the wrist (figure 5.5). A **capsular ligament** runs between the distal ends of the ulna and radius to the proximal row of

greater multangular or **multangulus major**), the trapezoid (also known as the **lesser multangular** or **multangulus minor**), the **capitate**, and the **hamate**.

Hands on . . . Downward pressure on the spot illustrated in figure 5.2 will bring you in contact with the hamate bone in the distal row of carpal bones.

There are five bones of the hand known as the metacarpal bones. The metacarpal bone of the thumb is usually referred to as the first metacarpal. The second metacarpal is that of the index finger, the third metacarpal is that of the middle finger, the fourth metacarpal is that of the ring finger, and the fifth metacarpal bone is that of the little finger. Distal to each of the metacarpal bones of the hand are the **phalanges** of the fingers. The thumb has two phalanges known as the proximal phalanx and the distal phalanx. The other fingers each consist of three phalanges: a proximal phalanx, a middle phalanx, and a distal phalanx.

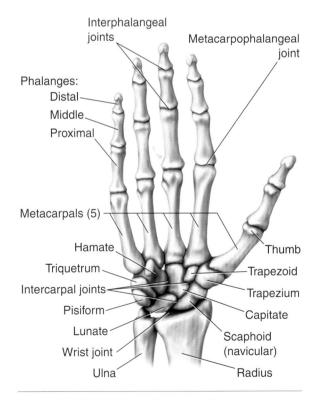

Figure 5.4 The wrist and intercarpal joints.

carpal bones. There is a **volar** (palmar) **radiocarpal ligament** and a **dorsal radiocarpal ligament.** The volar radiocarpal ligament is found between the anterior surface of the radius and its styloid process and the proximal row of carpal bones (figure 5.5). The dorsal radiocarpal ligament is found between the distal end of the radius and the proximal row of carpal bones (figure 5.5). The two additional ligaments of the wrist are collateral ligaments: the **radial** (lateral) and **ulnar** (medial) **collateral ligaments** (figure 5.5). The radial collateral ligament of the wrist runs between the styloid process of the radius and the scaphoid carpal bone. The ulnar collateral ligament runs between the styloid process of the ulna and the medial portions of the pisiform and triquetrum bones.

The **intercarpal joints** between the carpal bones of the wrist are connected by three forms of **intercarpal ligaments:** those that connect the four carpal bones in the proximal row, those that connect the four carpal bones in the distal row, and those that connect the carpals of the proximal row to those of the distal row. The intercarpal ligaments can be further divided into volar, dorsal (see figure 5.5), interosseous, radial and ulnar collateral, pisohamate, and pisometacarpal ligaments. All of these ligaments are referred to in this text simply as the intercarpal ligaments. The intercarpal joints move in a gliding motion.

The last group of joints that are considered to be part of the wrist are the carpometacarpal joints. There are five carpometacarpal joints: four between the four carpal bones in the distal carpal row and the bases of the four metacarpal bones of the hand, and one between the trapezium and the base of the first (thumb) metacarpal bone. The motion of these joints is gliding. The ligaments of the four carpometacarpal joints of the hand are the **dorsal, volar, interosseous,** and **capsular carpometacarpal ligaments.** The dorsal and volar (palmar)

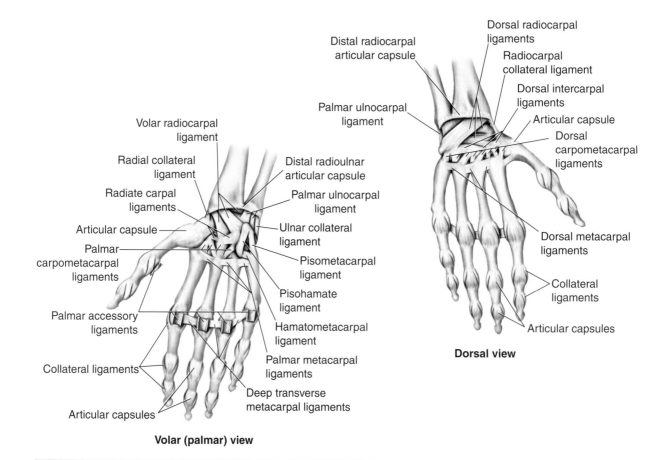

Volar (palmar) view

Dorsal view

Figure 5.5   The capsular ligament between the distal ends of the ulna and radius; the volar radiocarpal ligament; the dorsal radiocarpal ligament; the radial and ulnar collateral ligaments of the wrist; and the capsular and collateral ligaments of the interphalangeal joints.

carpometacarpal ligaments (see figure 5.5) are found between the dorsal and volar surfaces of the distal row of carpal bones and the bases of the metacarpal bones. The interosseous ligaments are found between the hamate and capitate bones and the bases of the third and fourth metacarpal bones. The capsular ligaments are located between the distal row of carpals and the bases of the four metacarpals of the hand. The first (thumb) carpometacarpal joint is unique compared with the four other carpometacarpal joints; its ligamentous structure consists of a loose capsular carpometacarpal ligament that is found between the trapezium and the base of the first (thumb) metacarpal bone. The carpometacarpal joint of the thumb is referred to as a saddle joint because of its shape.

Two additional ligamentous structures of the wrist are the flexor (volar) and extensor (dorsal) retinacula (figure 5.6), which are bands of connective tissue over the volar surface **(flexor retinaculum)** and the dorsal surface **(extensor retinaculum)** of the wrist. The flexor retinaculum forms a bridge over the carpal bones to form the carpal tunnel through which the flexor muscle tendons of the wrist and hand pass. There is a much smaller space between the extensor retinaculum and the carpal bones, through which pass the extensor tendons of the wrist and hand.

The joints of the hand and fingers consist of five metacarpophalangeal (MP) joints, articulations between the five long metacarpal bones of the hand and the five proximal phalanges of the fingers. The four fingers, having three phalanges each, also have **proximal interphalangeal (PIP) joints** and **distal interphalangeal (DIP) joints.** The thumb, having only two phalanges, has only one interphalangeal (IP) joint. All five MP, all four PIP, and all four DIP joints of the fingers and the IP joint of the thumb have capsular ligaments and ulnar and radial collateral ligaments (see figure 5.5).

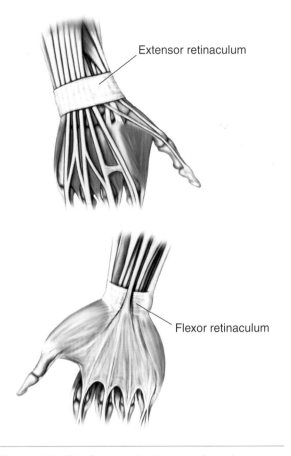

Extensor retinaculum

Flexor retinaculum

Figure 5.6   The flexor and extensor retinacula.

Focus on . . . CARPAL TUNNEL SYNDROME

In addition to the various tendons, blood vessels and nerves also pass through the carpal tunnel. Anything causing inflammation of these tissues in the tunnel, such as direct trauma or overuse of the muscles that have tendons passing through the tunnel, can cause swelling within the tunnel. Pressure on the tendons, blood vessels, or nerves can result in pain and diminished functioning of any of these tissues. This condition is often referred to as carpal tunnel syndrome.

Repetitive motions that stress the wrist may lead to carpal tunnel syndrome. A common cause of carpal tunnel syndrome is typing at a computer keyboard. In an effort to alleviate this stress, design engineers and biomechanists use an understanding of anatomy to produce new and safer keyboards. The field of study that attempts to improve biomechanical working conditions is known as ergonomics.

Movements of the Wrist and Hand

Remember that the primary movement among all the numerous joints of the wrist is defined as gliding. A combination of these gliding joint actions results in the wrist having four fundamental movements (figures 5.7 and 5.8): flexion (which takes place primarily in the radiocarpal joints), extension (which takes place primarily in the midcarpal joints), **ulnar deviation** (adduction), and **radial deviation** (abduction). Although movement at the wrist that results in the hand moving toward the midline of the body might be considered adduction, it is known as ulnar deviation. Likewise, moving the hand away from the midline of the body might be considered abduction, but it is known as radial deviation. Keep in mind that the wrist is a biaxial joint (i.e., it can move in two planes about two axes) and is capable of circumduction. Circumduction was defined in chapter 2 as a combination of fundamental movements of a biaxial or triaxial joint. The MP, PIP, DIP, and IP joints of the fingers and thumb all are capable of flexion and extension. The MP joints are also

Figure 5.8 Radial and ulnar deviation.

capable of abduction and adduction. The thumb is capable of additional movements that are presented later when the muscles of the thumb are discussed.

Wrist and Extrinsic Hand Muscles

Several muscles that are responsible for movement in the wrist actually originate above the elbow joint on either the medial or lateral epicondyle of the humerus. Although these muscles cross the elbow joint, they are not typically considered muscles that create movement in the elbow joint. They are considered muscles of the wrist and hand. Muscles that originate externally to the hand (on the humerus, ulnar, or radius) and insert within the hand are referred to as **extrinsic muscles** of the hand.

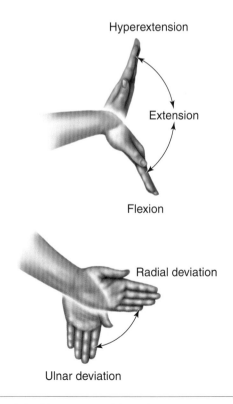

Figure 5.7 Movements of the wrist.

Anterior Muscles

There are five major muscles of the hand and wrist that appear on the anterior (volar) surface of the forearm. Four of these muscles (figure 5.9) originate on the medial epicondyle of the humerus on a structure known as the **common flexor tendon:** the **flexor carpi radialis,** the **flexor carpi ulnaris,** the **flexor digitorum superficialis,** and the **palmaris longus.** The fifth muscle, not part of the common flexor tendon, is the **flexor digitorum profundus.**

> **Flexor carpi radialis:** Part of the group of muscles originating from the medial epicondyle of the humerus off of the structure known as the common flexor tendon, this muscle inserts on the bases of the second and third metacarpal bones of the hand (figure 5.9). Contraction of the flexor carpi radialis muscle produces flexion and radial deviation of the wrist.

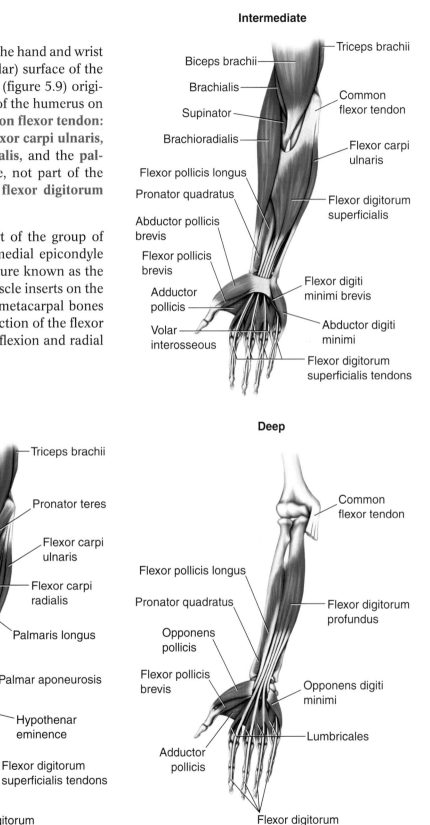

Figure 5.9 Superficial, intermediate, and deep muscles of the wrist and hand, anterior view.

Hands on . . . Make a fist and flex your wrist, find the tendon of the flexor carpi radialis, and trace it upward with your finger until you feel the belly of the muscle (figure 5.10).

Figure 5.11   Finding the flexor carpi ulnaris.

bases of the middle phalanges of the four fingers (figure 5.9). This muscle flexes the wrist, the MP joint, and the PIP joint of the four fingers.

Hands on . . . Apply pressure to the area of the forearm containing the tendon of the flexor digitorum superficialis and flex and extend your fingers. You should feel or see movement. This movement is produced by the flexor digitorum superficialis (figure 5.12).

> **Palmaris longus:** This is the fourth, and last, muscle (figure 5.9) that has its origin on the common flexor tendon, and it inserts on the palmar aponeurosis (figure 5.9). This muscle's very long tendon passes above the flexor retinaculum, unlike the tendons of the other common flexor tendon muscles, which pass beneath the retinaculum and through the carpal tunnel. Contraction of this muscle results in flexion of the wrist and tightening of the palmar fascia. The palmaris

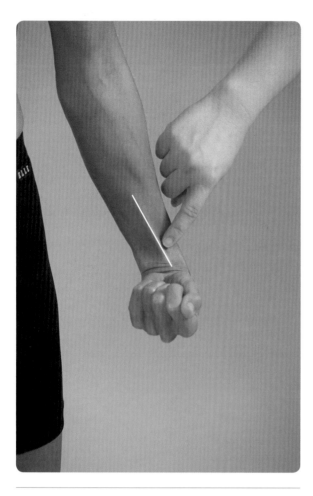

Figure 5.10   Finding the flexor carpi radialis.

> **Flexor carpi ulnaris:** Originating on the common flexor tendon, the flexor carpi ulnaris inserts on the pisiform and hamate carpal bones of the wrist and the base of the fifth metacarpal bone of the hand (figure 5.9). This muscle produces flexion and ulnar deviation of the wrist.

Hands on . . . Flex your wrist against a resistance, and note the tendon of the flexor carpi ulnaris (figure 5.11).

> **Flexor digitorum superficialis:** This muscle originates on the common flexor tendon; its tendon of insertion splits into four separate tendons that split again and insert on each side of the

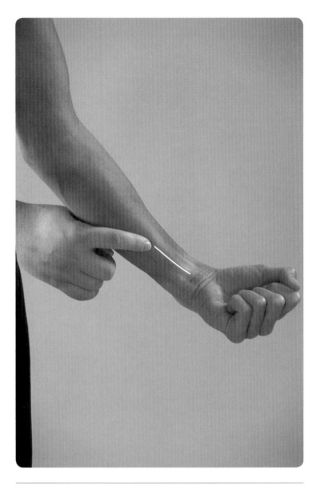

Figure 5.12 Finding the flexor digitorum superficialis.

Figure 5.13 Locating the palmaris longus.

longus is absent in 20% of people, but this is not a problem because its function is to assist the other flexor muscles and not to act as a prime flexor. The tendon of this muscle is often used to reinforce elbow ligaments that need surgical reconstruction.

Hands on . . . Press your fingers and thumb together, flex your wrist, and feel for movement in the lower third of the anterior aspect of your forearm (figure 5.13). The palmaris longus muscle creates this movement.

> **Flexor digitorum profundus:** This muscle is not part of the common flexor tendon group. It originates on the proximal portion of the volar surface of the ulna and divides into four tendons that pass through the carpal tunnel and split to insert on either side of the bases of the distal phalanges of the four fingers (figure 5.9). This muscle

is involved in flexion of the wrist, the four MP joints, the four PIP joints, and the four DIP joints of the fingers.

Posterior Muscles

Six major muscles of the hand and wrist appear on the posterior (dorsal) surface of the forearm (figure 5.14). Four of these muscles originate on the lateral epicondyle of the humerus on a structure known as the **common extensor tendon:** the **extensor carpi radialis brevis,** the **extensor carpi ulnaris,** the **extensor digitorum (communis),** and the **extensor digiti minimi (proprius).** The fifth and sixth muscles, not part of the common extensor tendon, are the **extensor carpi radialis longus** and the **extensor indicis.**

> **Extensor carpi radialis brevis:** This muscle (see figure 5.14) is one of four muscles originating

Superficial

Deep

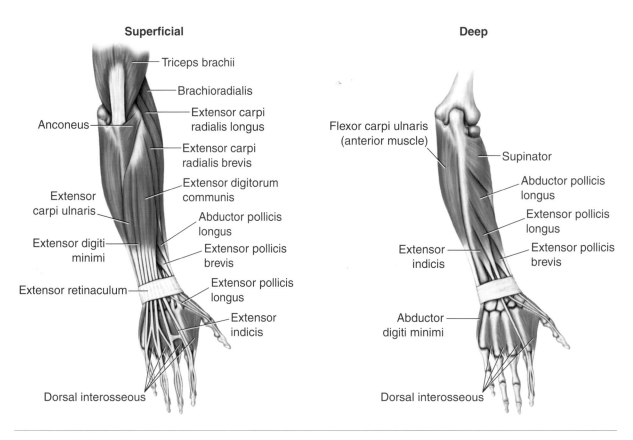

Figure 5.14   Superficial and deep muscles of the wrist and hand, posterior view.

from the common extensor tendon on the lateral epicondyle of the humerus and inserts on the dorsal aspect of the third metacarpal bone. The action produced by this muscle is extension and radial deviation of the wrist.

Hands on . . . Against resistance, extend your wrist. On the radial side of the dorsal aspect of the lower third of the forearm, you should see and feel two tendons: the extensor carpi radialis longus and the extensor carpi radialis brevis (figure 5.15).

> **Extensor carpi ulnaris:** Another of the four muscles originating on the common extensor tendon, this muscle (figure 5.14) inserts on the dorsal aspect of the fifth metacarpal bone, and it extends and ulnarly deviates the wrist.

Hands on . . . Extend your wrist while it is in ulnar deviation, and you will feel the extensor carpi ulnaris muscle (figure 5.16).

> **Extensor digitorum (communis):** The third muscle originating from the common extensor

Figure 5.15   Locating the extensor carpi radialis.

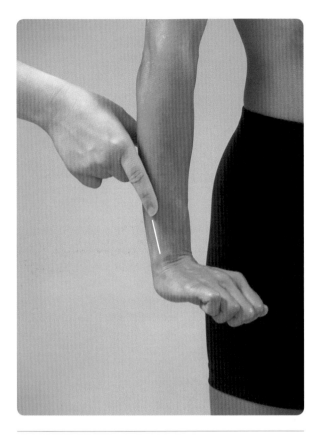

Figure 5.16 Locating the extensor carpi ulnaris.

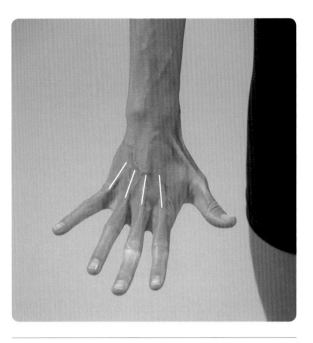

Figure 5.17 Viewing the extensor digitorum communis.

Hands on . . . Extend your little finger and feel for movement by placing pressure over the fifth metacarpophalangeal joint. This movement is created by the extensor digiti minimi (figure 5.18).

tendon, this muscle is typically identified as the extensor digitorum, although some texts add the additional term *communis* to the name to distinguish it from similarly named muscles in the foot. The muscle divides into four tendons and inserts on the bases of the distal phalanges of the four fingers (figure 5.14). The muscle extends the wrist, the MP joints, the PIP joints, and the DIP joints of all four fingers.

Hands on . . . Extend your wrist and flex and extend your fingers. Observe movement of the four tendons of the extensor digitorum on the dorsal aspect of your hand (figure 5.17).

> **Extensor digiti minimi (proprius):** The fourth muscle originating from the common extensor tendon sometimes includes the term *proprius*. Most texts simply call the muscle the extensor digiti minimi, indicating that the muscle extends the little finger (digiti minimi). The muscle inserts on the base of the proximal phalanx of the fifth finger and extends the wrist and the MP joint of the fifth finger (figure 5.14).

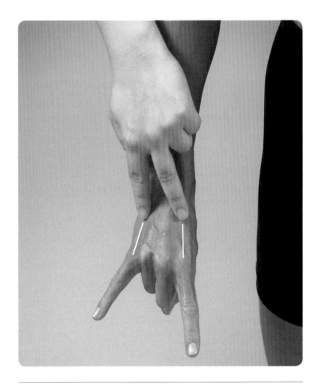

Figure 5.18 Locating the extensor indicis and extensor digiti minimi.

> **Extensor carpi radialis longus:** This is not a muscle of the common extensor tendon. It originates on the lateral supracondylar ridge of the humerus and inserts on the lateral side of the base of the second metacarpal bone. Its actions include extension and radial deviation of the wrist.

Hands on . . . Against resistance, extend your wrist. On the radial side of the dorsal aspect of the lower third of the forearm, you should see and feel two tendons: the extensor carpi radialis longus and the extensor carpi radialis brevis (see figure 5.15).

> **Extensor indicis:** This muscle originates on the dorsal surface of the distal portion of the ulna and inserts on the base of the proximal phalanx of the index finger (figure 5.14). Its primary function is to extend the MP joint of the index finger, and it also assists in extension of the wrist.

Hands on . . . Place pressure on your second MP joint, and feel for movement as you flex and extend your index finger. The movement is created by the tendons of the extensor digitorum and, just medial to it, the extensor indicis (see figure 5.18).

Intrinsic Muscles of the Hand

Whereas the preceding extrinsic muscles of the hand originate outside the hand and insert within the hand, the following muscles are the intrinsic muscles of the hand, originating and inserting totally within the hand. Three of these muscles (figure 5.9) make up an anatomical structure known as the **hypothenar eminence** (the **abductor digiti minimi,** the **flexor digiti minimi brevis,** and the **opponens digiti minimi**). *Hypo* (from Greek origin, meaning "less than") indicates that these are a group of muscles involved with movement of the fifth (little) finger.

> **Abductor digiti minimi:** The first of three muscles making up the hypothenar eminence, this muscle originates on the pisiform bone of the wrist and inserts on the proximal phalanx of the little finger. It both abducts and assists with flexion of the fifth MP joint.

> **Flexor digiti minimi brevis:** The second muscle of the hypothenar eminence originates

on the hamate bone of the wrist and inserts on the proximal phalanx of the fifth finger. It assists in flexion of the fifth MP joint.

> **Opponens digiti minimi:** This is the third and last muscle making up the hypothenar eminence. It originates on the hamate bone of the wrist and inserts on the fifth metacarpal and proximal phalanx bone of the fifth finger, and it assists with flexion and adduction of the fifth MP joint. The ability to bring the little finger toward the thumb to allow grasping is known as **opposition.**

A number of muscles of the hand are not involved with movement of the fifth finger but are involved in the movement of other finger joints. These muscles appear as three distinct muscle groups: the **dorsal** and **volar (palmar) interossei** (figure 5.19) and the **lumbricales.**

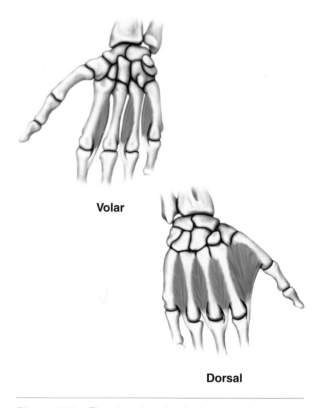

Volar

Dorsal

Figure 5.19   The dorsal and volar interossei muscles of the hand.

> **Dorsal interossei:** There are four dorsal interosseous (between the bones) muscles that originate on the four metacarpal bones of the hand (index, middle, ring, and little fingers) and insert on the second (index), third (middle), and fourth (ring)

proximal phalanges. Two of these muscles attach to the lateral aspect of the second and third phalanges, whereas the other two attach to the medial aspect of the third and fourth phalanges. The first interosseus can be easily located on the dorsal side of the hand (figure 5.20).

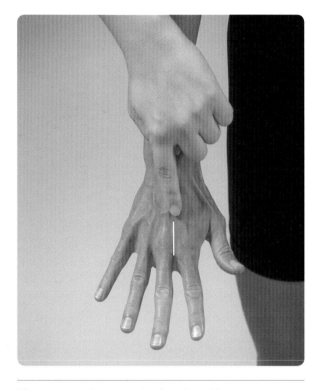

Figure 5.20   Locating the first dorsal interosseus.

Hands on . . . Spread your fingers and observe what the second, third, and fourth MP joints do. The dorsal interosseous muscles cause the second MP joint to abduct (move laterally from the midline of the hand) and the fourth MP joint to adduct (move medially from the midline of the hand), whereas the third MP joint remains stationary. Why? Depending on the angle of the MP joints, these muscles also assist with flexion and extension of the joints.

> **Palmar interossei:** There are three palmar (volar) interosseous muscles that originate on the second, fourth, and fifth metacarpal bones of the hand and insert on the second, fourth, and fifth proximal phalanges. Two of these muscles attach to the medial aspects of the fourth and fifth phalanges, whereas the third attaches to the lateral aspect of the second phalanx.

Hands on . . . Spread your fingers. As you return your fingers toward the middle finger, observe what the second, fourth, and fifth MP joints do. The palmar interosseous muscles cause the second MP joint to adduct (move medially toward the midline of the hand) and the fourth and fifth MP joints to abduct (move laterally to the midline of the hand), whereas the third MP joint remains stationary. Why? Depending on the angle of the MP joints, these muscles also assist with flexion and extension of the joints.

> **Lumbricales:** These four muscles are found deep within the hand; they originate on the tendons of the flexor digitorum profundus and insert on the tendons of the extensor digitorum communis in the area of the proximal phalanges (figure 5.9). This group of muscles assists with the flexion of the MP joints and extension of the PIP and DIP joints (figure 5.21).

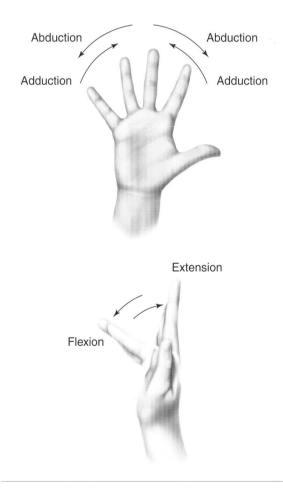

Figure 5.21   Movements of the metacarpophalangeal joints.

Muscles of the Thumb

Muscles involved with movement of the thumb are divided into extrinsic (originating outside of and inserting within the hand) and intrinsic (originating and inserting within the hand) groups.

Extrinsic Muscles

These muscles all originate extrinsically, proximal to the thumb, and insert within the thumb. Note that all contain the word *pollicis* (from the Latin word *pollex*, for thumb) in their names.

> **Extensor pollicis longus:** This muscle originates on the ulna, inserts on the distal phalanx of the thumb, and extends both the IP and MP joints of the thumb (figure 5.14). This muscle also assists with radial deviation of the wrist and supination of the forearm.

> **Extensor pollicis brevis:** This muscle originates on the radius and inserts on the proximal phalanx of the thumb (figure 5.14). It extends the MP joint and assists with radial deviation of the wrist.

The tendons of the extensor pollicis brevis and the **abductor pollicis longus,** as they cross the wrist, form a depression (fossa) directly over the scaphoid carpal bone. This depression is commonly known as the **anatomical snuffbox** (figure 5.22). This term dates to the period in history when tobacco in the form of snuff was used by placing it on this area and then inhaling it through the nose.

> **Abductor pollicis longus:** This muscle originates on the ulna and inserts on the base of the first metacarpal bone (figure 5.14). It abducts the first metacarpal bone and also assists with radial deviation and flexion of the wrist.

Hands on . . . Abduct your thumb, and apply pressure at the base of the thumb between the two tendons (see figure 5.22). This muscle is difficult to palpate on some individuals; try several people until you can palpate this muscle as your subject abducts the thumb (figure 5.23).

> **Flexor pollicis longus:** This muscle originates on the radius and inserts on the distal phalanx of the thumb (see figure 5.9). It flexes both the IP and MP joints of the thumb and also assists with flexion of the wrist (figure 5.24).

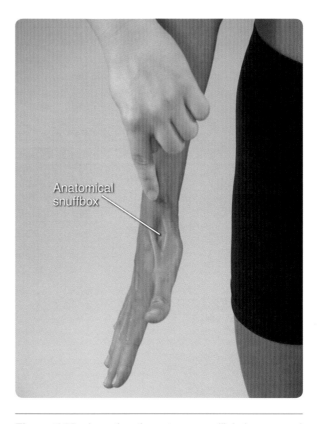

Anatomical snuffbox

Figure 5.22   Locating the extensor pollicis longus and brevis.

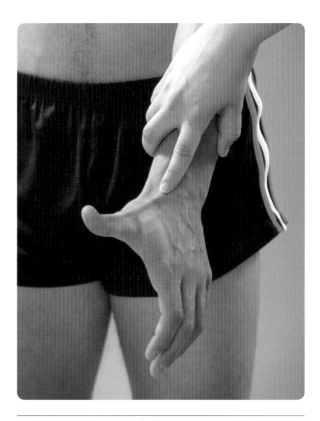

Figure 5.23   Locating the abductor pollicis longus.

Figure 5.24 Locating the flexor pollicis longus.

Intrinsic Muscles

The following group of muscles both originate and insert totally within (intrinsically to) the hand. The bellies of these muscles form a thick pad of tissue just proximal to the thumb that is commonly known as the **thenar eminence** (see figure 5.9). Again, note that all four muscles have the term *pollicis* (thumb) in their names.

> **Abductor pollicis brevis:** Originating from the trapezium and scaphoid bones and inserting on the base of the proximal phalanx of the thumb, this muscle assists with abduction of the thumb (figure 5.9).

Hands on . . . Abduct your thumb and observe or feel the movement on the lateral aspect of the thenar eminence. This movement is produced by the abductor pollicis brevis.

> **Flexor pollicis brevis:** This muscle (figure 5.9) originates on the trapezium bone and inserts on the proximal phalanx of the thumb. It assists the flexor pollicis longus in flexing the MP joint of the thumb.

Hands on . . . Flex your thumb toward your second finger and feel the flexor pollicis brevis move (figure 5.25).

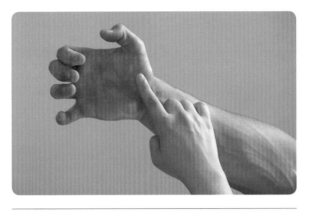

Figure 5.25 Finding the flexor pollicis brevis.

> **Opponens pollicis:** Originating on the trapezium and inserting on the first metacarpal bone, this muscle adducts (opposes) the thumb (figure 5.9).

Hands on . . . Move your thumb toward the four fingers and palpate along the lateral aspect of the thumb's MP joint. The movement you feel is created by the opponens pollicis (figure 5.26).

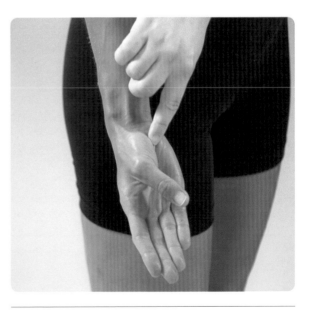

Figure 5.26 Locating the opponens pollicis.

> **Adductor pollicis:** This two-headed muscle has one head (oblique) originating on the capitate and second and third metacarpal bones and the other head (transverse) originating on the third metacarpal bone. The two heads combine and insert on the base of the thumb's proximal phalanx (figure 5.9). The action of this muscle is adduction (opposition) of the thumb.

Hands on . . . Apply the thumb and index finger of your opposite hand to the space between the thumb and index finger, feeling for the tendon of the adductor pollicis.

Figures 5.27 through 5.29 illustrate the movements that the thumb is capable of performing as the result of actions by its extrinsic and intrinsic muscles. The thumb is capable of flexion and

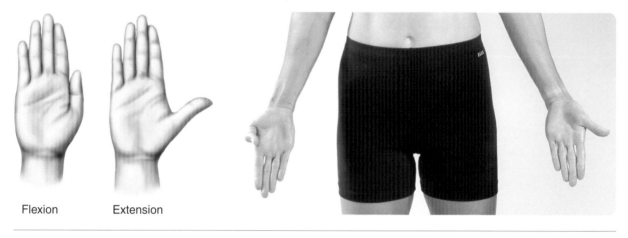

Flexion   Extension

Figure 5.27   Flexion and extension of the thumb.

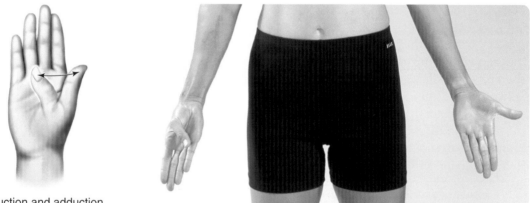

Abduction and adduction

Figure 5.28   Abduction and adduction of the thumb.

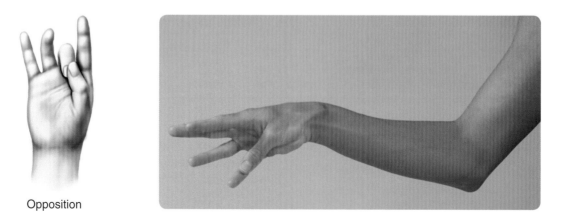

Opposition

Figure 5.29   Opposition of the thumb.

extension (figure 5.27), adduction and abduction (figure 5.28), and movement in opposition (figure 5.29). Note that, in describing thumb movements in the anatomical position, there are exceptions to the usual references. Thumb flexion and extension appear to take place in the frontal plane, whereas abduction and adduction appear to take place in the sagittal plane.

REVIEW OF TERMINOLOGY

The following terms were discussed in this chapter. Define or describe each term, and where appropriate, identify the location of the named structure either on your body or in an appropriate illustration.

abductor digiti minimi
abductor pollicis brevis
abductor pollicis longus
adductor pollicis
anatomical snuffbox
capitate
capsular carpometacarpal ligament
capsular ligament
carpals
carpometacarpal joints
common extensor tendon
common flexor tendon
distal interphalangeal joints
dorsal carpometacarpal ligament
dorsal interossei
dorsal radiocarpal ligament
extensor carpi radialis brevis
extensor carpi radialis longus
extensor carpi ulnaris
extensor digiti minimi (proprius)
extensor digitorum (communis)
extensor indicis
extensor pollicis brevis
extensor pollicis longus

extensor retinaculum
extrinsic muscles
flexor carpi radialis
flexor carpi ulnaris
flexor digiti minimi brevis
flexor digitorum profundus
flexor digitorum superficialis
flexor pollicis brevis
flexor pollicis longus
flexor retinaculum
greater multangular
hamate
hypothenar eminence
intercarpal joints
intercarpal ligaments
interosseous carpometacarpal ligament
lesser multangular
lumbricales
lunate
metacarpals
metacarpophalangeal joints
midcarpal joints
multangulus major

multangulus minor
navicular
opponens digiti minimi
opponens pollicis
opposition
palmar interossei
palmaris longus
phalanges
pisiform
prehensile hands
proximal interphalangeal joints
radial collateral ligament
radial deviation
radiocarpal joints
scaphoid
thenar eminence
trapezium
trapezoid
triquetrum
ulnar collateral ligament
ulnar deviation
volar carpometacarpal ligament
volar radiocarpal ligament

SUGGESTED LEARNING ACTIVITIES

1. Extend and abduct the thumb on one of your hands. Note the two tendons that rise up at the proximal end of the thumb where it joins the wrist. With your other thumb, apply pressure downward between the two tendons.

   a. Name the two tendons.
   b. Give the common name for the area where you are applying pressure.
   c. What carpal bone lies directly beneath the area where you are applying pressure?

2. Flex both of your wrists as much as you can and then extend both of your wrists as much as you can. Do this from the anatomical position. Repeat both motions, only this time start with your hands in tight fists.

   a. Which motion, from the anatomical position, had greater range of motion: flexion or extension?

   b. What happened to the range of motion for both flexion and extension when the exercise was performed while you made a fist? If there was any change in the range of motion of the wrist between the neutral position and when a fist was made, what anatomical structures were responsible for these changes?

3. Holding your hand in the anatomical position, spread your fingers apart. What movements occurred at the

   a. index finger MP joint (by what muscle)?
   b. middle finger MP joint (by what muscle)?
   c. ring finger MP joint (by what muscle)?
   d. little finger MP joint (by what muscle)?

(continued)

---

4. With your fingers spread apart, return them to the anatomical position. What movements occurred at the

   a. index finger MP joint (by what muscle)?
   b. middle finger MP joint (by what muscle)?
   c. ring finger MP joint (by what muscle)?
   d. little finger MP joint (by what muscle)?

5. Touch the tip of your thumb to the tip of your little finger.

   a. What movements occurred in the thumb joints?
   b. What muscles performed these movements?

MULTIPLE-CHOICE QUESTIONS

1. Which of the following joints is not capable of circumduction?

   a. distal interphalangeal
   b. metacarpophalangeal
   c. wrist
   d. glenohumeral

2. Movement of the wrist into abduction is anatomically known as

   a. ulnar deviation
   b. radial deviation
   c. flexion
   d. extension

3. Which one of the following carpal bones is in the distal row of carpals?

   a. hamate
   b. scaphoid
   c. pisiform
   d. lunate

4. The structure known as the anatomical snuffbox is formed by the tendons of the extensor pollicis brevis and the

   a. flexor pollicis longus
   b. abductor pollicis longus
   c. extensor pollicis longus
   d. adductor pollicis longus

5. The lumbricales, located deep within the hand, both extend the interphalangeal joints and flex the

   a. metacarpophalangeal joints
   b. intercarpal joints
   c. carpometacarpal joints
   d. midcarpal joints

6. Which of the following muscles does not originate on the common extensor tendon?

   a. extensor carpi ulnaris
   b. extensor carpi radialis longus
   c. extensor digitorum
   d. extensor carpi radialis brevis

7. The muscles of the forearm that flex and extend only the wrist insert on either the carpal bones or the

   a. metacarpals
   b. proximal phalanges
   c. middle phalanges
   d. distal phalanges

8. The hypothenar eminence is formed by muscles that move the

   a. thumb
   b. index finger
   c. little finger
   d. ring finger

9. Which of the following carpal bones is also known as the navicular bone?

   a. pisiform
   b. scaphoid
   c. hamate
   d. lunate

10. Which of the following muscles does not originate on the common flexor tendon?

    a. flexor carpi ulnaris
    b. flexor digitorum superficialis
    c. flexor digitorum profundus
    d. flexor carpi radialis

11. Which of the following muscles is considered an extrinsic muscle of the little finger?

    a. extensor digiti minimi
    b. flexor digiti minimi brevis
    c. abductor digiti minimi
    d. opponens digiti minimi

12. Which of the following joints is considered biaxial?

    a. interphalangeal
    b. proximal interphalangeal
    c. metacarpophalangeal
    d. distal interphalangeal

13. Which of the following muscles is not considered an intrinsic muscle of the thumb (thenar eminence)?

    a. flexor pollicis brevis
    b. abductor pollicis brevis
    c. flexor pollicis longus
    d. adductor pollicis

14. The opponens pollicis muscle moves the thumb

    a. toward the little finger
    b. toward the radius
    c. into flexion
    d. into extension

15. Which of the following movements is not considered a fundamental movement of the wrist?

    a. flexion
    b. ulnar deviation
    c. radial deviation
    d. circumduction

FILL-IN-THE-BLANK QUESTIONS

1. The only flexor muscle of the wrist whose tendon of insertion does not pass under the flexor (volar) retinaculum is the _____.

2. Extension of the wrist occurs primarily at the _____ joints.

3. The tendons passing through the structure known as the carpal tunnel are considered primarily _____ of the wrist and hand.

4. The thenar eminence is formed by the intrinsic muscles of the _____.

5. Flexion of the wrist primarily occurs at the _____ _____ joints.

6. Ulnar deviation of the wrist in the frontal plane is otherwise also known as _____ of the wrist.

7. The extensor indicis extends the MP, PIP, and DIP joints of the _____ finger.

8. The most lateral carpal bone of the proximal carpal row is the _____.

9. The movement of the thumb that allows us to grasp objects is known as _____.

10. The radiocarpal joint is the articulation between the radius, the lunate, and the _____ bone.

11. The band of tissue at the wrist that keeps the wrist and hand flexor muscles from rising up under tension is known as the _____.

12. The wrist joint is stabilized on the medial and lateral sides by _____ ligaments from the radius and ulna.

13. The joints between the bones of the wrist and the bones of the hand are known as the _____ _____ joints.

14. The bone lying directly beneath the anatomical snuffbox is the carpal bone known as the _____ _____.

CHAPTER SIX

Nerves and Blood Vessels of the Upper Extremity

The structures known as nerves, arteries, and veins were presented in chapter 1. These structures are important to movement because they provide the stimulus to contract (nerves), they provide the blood supply that carries nutrients (arteries), and they remove by-products of the muscle's efforts (veins) to move the bones. The nervous system and the vascular system are very complex and have various functions. The nerves and blood vessels pertinent to the musculature of each anatomical section of this text (upper extremity, spinal column and thorax, lower extremity) are presented after the bones, ligaments, and muscles have been discussed.

The **nerves** of the upper extremity originate in the spinal cord from the first cervical to the first thoracic vertebral section and are commonly classified as motor nerves, sensory nerves, or mixed nerves containing both motor and sensory capabilities. Motor nerves innervate muscle and, when stimulated voluntarily or involuntarily, cause muscle fibers to contract and create movement in the joints that these muscle fibers cross. These nerves are presented in the summary table at the end of the upper-extremity part of this text (after this chapter).

The major upper-extremity arteries supply the fuel essential for muscular contraction, and the major veins carry away the waste products of muscular contraction. The major arteries supplying muscles are also listed in the summary table at the end of this part.

It is the intent of this text to examine the anatomy of the nervous system and the vascular system only as those structures pertain to the musculature. Figure 6.1 shows some of the relevant nerves that are covered in this text.

Spinal nerves innervating the musculature form **plexuses,** or networks. In the upper extremity, the nerves arise from the **brachial plexus** (C5, C6, C7, C8, and T1 nerve roots and often the nerve roots from C4 and T2 also) (figure 6.2). Although there are only seven cervical vertebrae, there are eight cervical spinal nerve roots. This is easily understood when you realize that the first cervical nerve root originates *above* the first cervical vertebra and the eighth cervical nerve root originates *below* the seventh cervical vertebra.

Nerves of the Brachial Plexus

Spinal nerves have both anterior and posterior rami (branches or arms). The anterior rami form the brachial plexus (figure 6.3), which is typically divided into two parts based on position relative to the clavicle: the **supraclavicular** (above the clavicle) and the **infraclavicular** (below the clavicle) parts.

Supraclavicular Nerves

The **dorsal scapular nerve** (C5 anterior and posterior rami) innervates the levator scapulae and rhomboid major and minor muscles (figure 6.3). The **long thoracic nerve** (C5, C6, and C7 anterior and posterior rami) (figures 6.2 and 6.3) innervates the serratus anterior muscle. Not illustrated are nerves (C2–C8) to the scalene and longus colli muscles of the cervical spine and a communicating nerve (C5) to the phrenic nerve of the cervical plexus, which are discussed in chapter 7 on the spinal column. Two nerves coming from one of three groups of nerves of the brachial plexus known as the *upper, middle,* and *lower trunks* are the **subclavian nerve** (C4, C5, C6), innervating the subclavian muscle, and the **suprascapular nerve** (C4, C5, C6) (figures 6.2 and 6.3), innervating the supraspinatus and infraspinatus muscles of the rotator cuff.

Infraclavicular Nerves

Three large cords of nerves (lateral, medial, and posterior) are formed in the brachial plexus from divisions of the upper, middle, and lower trunks of spinal nerves (see figure 6.3). The infraclavicular nerves arising from the **lateral cord** are the **lateral anterior thoracic nerve** (C5, C6, C7), innervating the pectoralis major muscle; the **lateral aspect of the median nerve** (C5, C6, C7), innervating most of the anterior forearm muscles and some in the hand; and the **musculocutaneous nerve** (C4, C5, C6) (figure 6.4), innervating the anterior arm muscles.

Hands on . . . Applying pressure between the triceps brachii and the biceps brachii on the medial aspect of your arm may produce a tingling sensation in your hand because you are compressing the median nerve (figure 6.5).

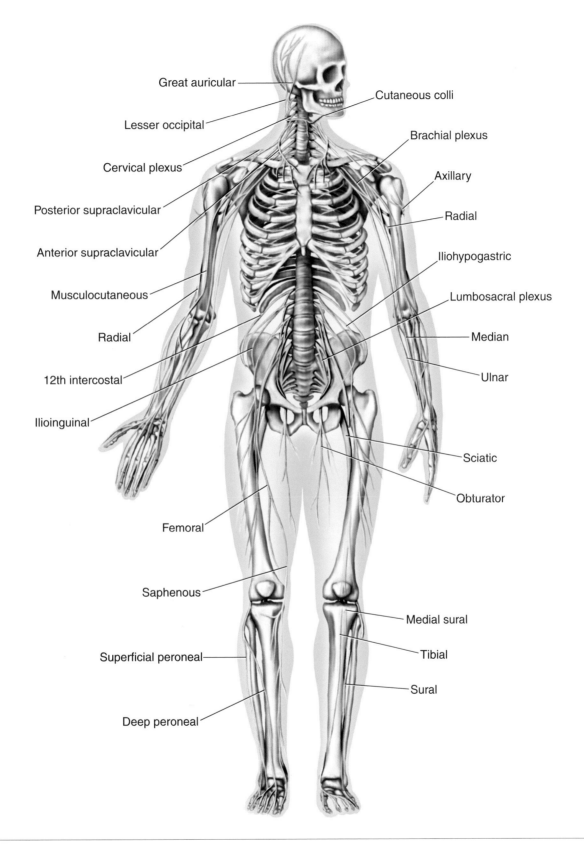

Figure 6.1 Nervous system superimposed over the skeleton.

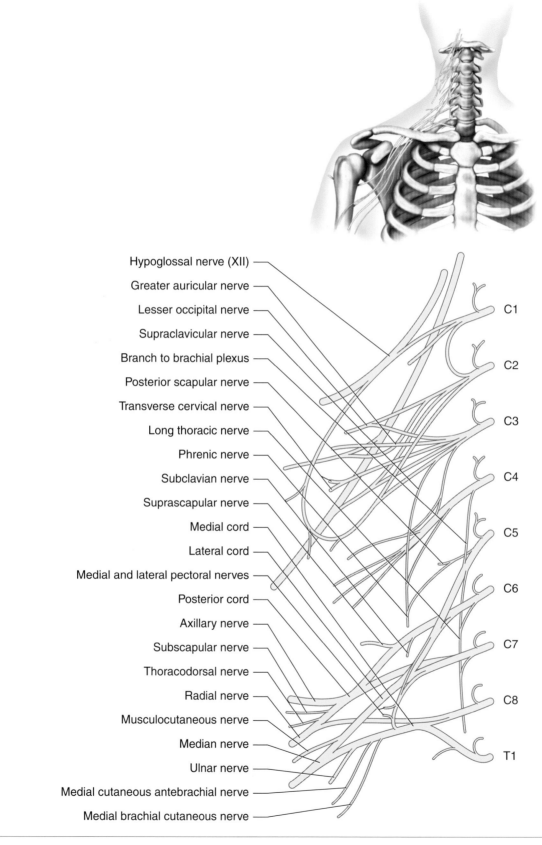

Hypoglossal nerve (XII)
Greater auricular nerve
Lesser occipital nerve
Supraclavicular nerve
Branch to brachial plexus
Posterior scapular nerve
Transverse cervical nerve
Long thoracic nerve
Phrenic nerve
Subclavian nerve
Suprascapular nerve
Medial cord
Lateral cord
Medial and lateral pectoral nerves
Posterior cord
Axillary nerve
Subscapular nerve
Thoracodorsal nerve
Radial nerve
Musculocutaneous nerve
Median nerve
Ulnar nerve
Medial cutaneous antebrachial nerve
Medial brachial cutaneous nerve

C1
C2
C3
C4
C5
C6
C7
C8
T1

Figure 6.2   View of C1 through T1 nerve roots.

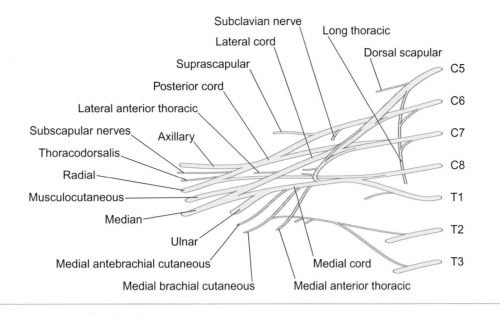

Subclavian nerve

Long thoracic

Lateral cord

Dorsal scapular

Suprascapular

C5

Posterior cord

C6

Lateral anterior thoracic

C7

Subscapular nerves   Axillary

C8

Thoracodorsalis

Radial

T1

Musculocutaneous

Median

T2

Ulnar

Medial antebrachial cutaneous

Medial cord   T3

Medial brachial cutaneous

Medial anterior thoracic

Figure 6.3   Isolated view of the brachial plexus.

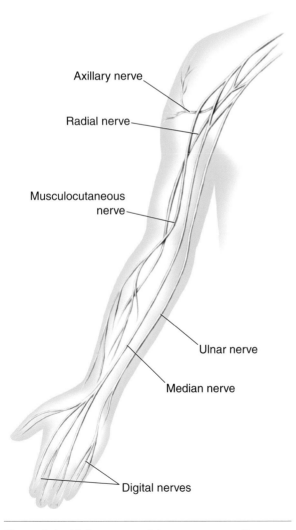

Axillary nerve

Radial nerve

Musculocutaneous
nerve

Ulnar nerve

Median nerve

Digital nerves

Figure 6.4   Major nerves of the upper extremity.

There are five major nerves of the brachial plexus that arise from the **medial cord** (see figure 6.3). The **medial anterior thoracic nerve** (C8, T1) innervates the pectoralis major and minor muscles (see figure 6.3). Joining the previously mentioned lateral aspect is the **medial aspect of the median nerve** (C8, T1), which innervates most of the anterior muscles of the forearm and some of the hand (see figures 6.4 and 6.5). The **medial cutaneous nerve of the forearm** (medial antebrachial cutaneous nerve) (T1) is *not* a motor nerve and innervates the skin and fascia on the medial surface of the arm. The **ulnar nerve** (C8, T1) innervates the flexor carpi ulnaris and the muscles of the medial aspect of the hand not innervated by the median nerve (see figure 6.4).

Hands on . . . The ulnar nerve is easily located between the medial epicondyle of the humerus and the olecranon process of the ulna (figure 6.6). Sudden pressure on this area often produces a tingling sensation in the forearm and is referred to as "hitting your funny bone."

The fifth nerve, the **medial brachial cutaneous nerve,** is not a motor nerve and innervates the skin and fascia on the medial aspect of the upper arm down to the medial epicondyle and olecranon process areas of the elbow.

Four brachial plexus nerves compose the **posterior cord** (see figure 6.3). The **axillary nerve** (C5, C6) (figure 6.4) innervates the deltoid and teres

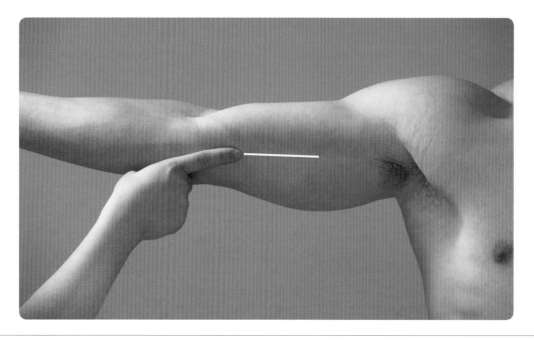

Figure 6.5   Locating the median nerve.

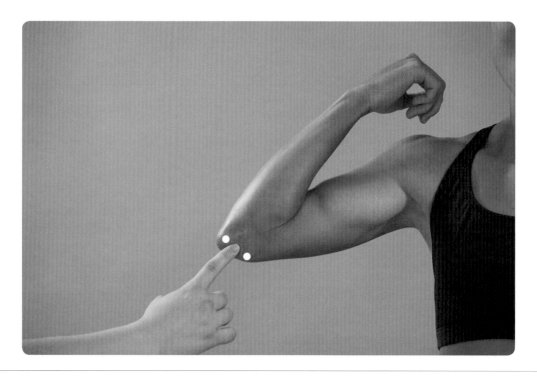

Figure 6.6   Finding the ulnar nerve.

minor muscles. The **radial nerve,** also known as the **musculospiral nerve** (C5, C6, C7, C8, T1), innervates the posterior muscles of the arm and forearm (see figure 6.4). The **subscapular nerves** (C5, C6), the **upper subscapular nerve** and the lower subscapular nerve, innervate the subscapularis and teres major muscles (figure 6.3). The **dorsal thoracic nerve,** also known as the **thoracodorsalis** (C6, C7, C8), innervates the **latissimus dorsi** muscle (see figure 6.3).

Focus on . . . VERTEBRAL TRAUMA

In a "pinched" nerve or a "stinger" type of trauma in the cervical region or upper extremity, the nerves of the brachial plexus are impinged or stretched. This type of trauma often occurs from the performance of questionable (if not illegal in football) techniques such as "butt-blocking" and "spear-tackling." Butt-blocking is an offensive player's technique that occurs when a blocker makes initial contact with an opponent by striking the opponent's chest ("in the numbers") with the blocker's helmet (face, head, forehead, nose). This can result in the blocker incurring a compression of the cervical vertebrae, causing the impingement of brachial plexus nerves (from a direct head-on blow) or causing the blocker's cervical spine to be excessively laterally flexed. In turn, either or both of the brachial plexus nerves are stretched on one side of the cervical spine while being impinged between cervical vertebrae on the opposite side (when the blow is struck off center causing the cervical spine to laterally flex). Spear-tackling (or spearing, headhunting) is a defensive player's technique in which a ballcarrier is struck by the tackler's head as the initial point of contact with the opponent. The same or similar results that could occur from butt-blocking are possible with this tackling technique. Aside from these football examples, any force that may cause compression of the cervical spine or excessive lateral flexion of the cervical spine can result in trauma, impinging or stretching the brachial plexus nerves or even both.

In order for players to avoid injury to the structures of the brachial plexus, coaches need to learn proper coaching techniques. In the event of injury to the brachial plexus, coaches and athletic trainers must have a sufficient background in first aid to minimize further injury. An understanding of anatomy is vital to the prevention and treatment of pinched nerves.

Major Arteries of the Upper Extremity

The blood vessels of the body are divided into **arteries, arterioles, capillaries, venules,** and **veins.** The arteries (see figure 1.22 on page 17), which carry blood away from the heart, and some of the larger arterioles are easily observed, whereas the smaller arterioles and capillaries are microscopic. The artery walls are lined with **smooth muscle fibers** that add additional pumping action; the veins have few or no smooth muscle fibers but have small valves that permit blood flow in only one direction. Any attempt to reverse this direction is blocked by the closing of these valves. The veins, which carry blood to the heart, and some of the larger venules are easily observed, whereas the smaller venules and capillaries are microscopic. Cells receive their nutrients and release their by-products at the capillary level.

The major arteries of the upper extremity include the **subclavian, axillary, brachial, radial, deep volar arch,** and **ulnar arteries** (figure 6.7). The subclavian artery originates near the heart, runs posterior to the clavicle, and has three major parts: the costocervical trunk, the internal mammary, and the thyrocervical trunk. Each of these three parts has branches that supply various structures of the thorax and upper extremity.

Hands on . . . Apply finger pressure (take your pulse) to the middle superior aspect of your clavicle to palpate the subclavian artery (figure 6.8). Note that using your fingers rather than your thumb is the preferred technique for taking a pulse. Some authorities believe that the thumb also possesses a pulse and could give a false impression when used to palpate for the presence of a pulse in another anatomical area.

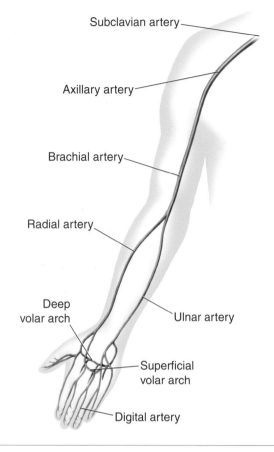

Figure 6.7   The major arteries of the upper extremity.

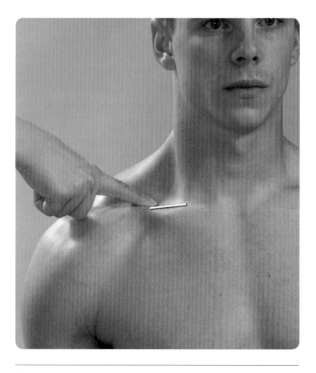

Figure 6.8   Locating the subclavian artery.

The axillary artery is an extension of the subclavian artery that begins at the outside border of the first rib and runs to the lower portion of the teres major muscle, where it becomes known as the brachial artery.

Hands on . . . Palpate the axillary area beneath the coracobrachialis muscle for a pulse (figure 6.9). This is a pulse from the axillary artery.

The brachial artery extends from the axilla to the elbow, where it divides into the radial and ulnar arteries. The branches of the brachial artery supply the structures of the upper arm (shoulder to elbow).

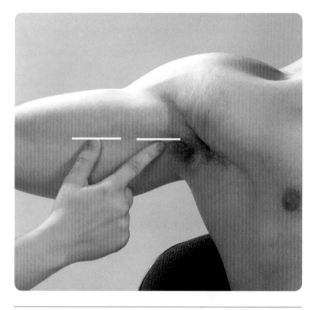

Figure 6.9   Locating the axillary and brachial arteries.

Hands on . . . Palpate the area on the medial aspect of the upper arm between the triceps brachii and the biceps brachii muscles (take your pulse) to locate the brachial artery (figure 6.9). This area is often considered a *pressure point* for attempting to reduce blood flow below that site (as in applying pressure in first aid for bleeding in the forearm or hand).

The radial artery (the lateral branch of the brachial artery) and its branches supply the anterior lateral structures from the elbow, through the forearm and wrist, into the hand. This artery is often associated with taking a pulse, because the radial artery is compressed against the distal end of the radius bone (figures 6.7 and 6.10). The medial branch of the brachial artery is the ulnar artery (figures 6.7 and 6.11). The ulnar artery and its branches supply the posterior and anterior medial structures of the forearm and hand. In the wrist, a branch of the ulnar artery (dorsal ulnar carpal branch) joins a branch of the radial artery (**dorsal radial carpal branch**) to form the structure known as the **dorsal carpal branch** (figure 6.12). In the palm of the hand (volar surface), the radial artery joins with branches of the ulnar artery to form two structures known as the **superficial volar arch** and the deep volar arch (figures 6.7 and 6.12). Branches from these volar arches, including the **digital arteries** on either side of the fingers, supply the structures of the fingers and thumb (see figures 6.7 and 6.12).

Figure 6.11 Locating the ulnar artery.

Figure 6.10 Finding the radial artery.

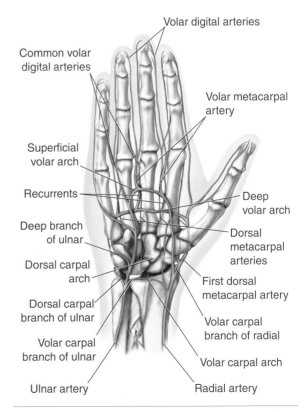

Figure 6.12 The dorsal carpal branch of the ulnar artery, the dorsal carpal arch, and the superficial and deep volar arches.

Major Veins of the Upper Extremity

The veins that return blood to the heart (see figure 1.23 on page 18) are commonly divided into **deep veins** and **superficial veins.** With a few exceptions, the deep veins have the same names as the arteries they parallel, such as the **subclavian vein** and **axillary vein.** The superficial veins have specific names and are located near the skin. Unlike the arteries, veins do not always appear exactly where one might expect, and often they may be absent altogether.

The major deep veins of the upper extremity (figure 6.13) are the subclavian and the axillary. The axillary vein drains the blood from the upper extremity into the subclavian vein, which combines with the **internal jugular vein** to form the **brachiocephalic vein** (figure 6.14). Feeding into the axillary vein are the veins paralleling the major arteries of the upper extremity: the **brachial vein, radial vein, ulnar vein, venous arch,** and **digital veins** (figure 6.13).

The major superficial veins of the upper extremity (figure 6.13) include the **basilic,** the **cephalic,** and the **median veins.** The **median cubital vein** crosses the anterior aspect of the elbow joint (cubital fossa) laterally to medially, and the cephalic vein can be observed superior to the brachioradialis muscle and slightly lateral to the biceps brachii muscle (figure 6.15). The basilic vein originates on the ulnar aspect of the venous arch in the hand. At the elbow, the basilic vein receives the median cubital vein and, on reaching the lower portion of the teres major muscle, joins the axillary vein (figure 6.16). It drains the structures on the ulnar side of the volar and dorsal aspects of the hand, forearm, and upper arm. The

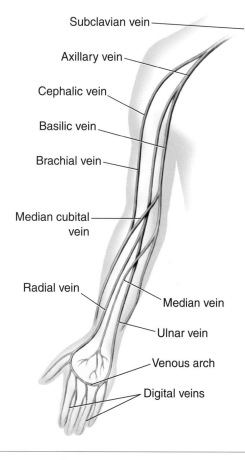

Figure 6.13   The major veins of the upper extremity.

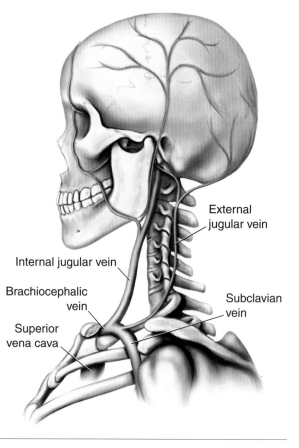

Figure 6.14   The brachiocephalic and internal jugular veins.

cephalic vein originates on the radial side of the dorsal venous arch in the hand; at the elbow it is connected to the basilic vein by the median cubital vein and then joins the axillary vein. Also at the elbow, the cephalic vein connects to the **accessory cephalic vein** (figure 6.17). The cephalic vein drains the structures from the lateral aspect of the hand, forearm, and upper arm. The median vein originates in the **palmar venous vessels** of the hand and connects to the **basilic cubital vein.** In the elbow area, the median vein is joined by the median cubital vein, which drains the medial aspect of the forearm and hand (figure 6.18). The median cubital vein is often used when blood is donated or a sample of blood is needed for testing purposes.

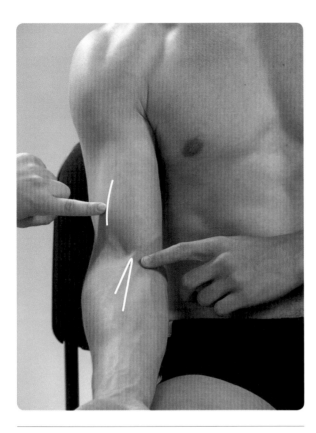

Figure 6.15   Locating the median cubital and cephalic veins.

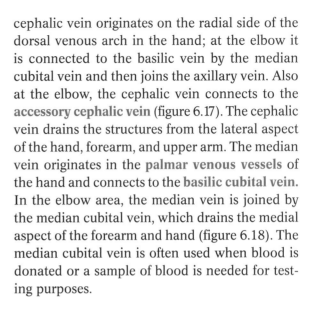

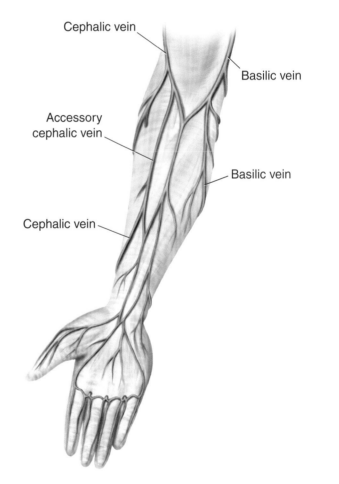

Cephalic vein

Basilic vein

Accessory cephalic vein

Basilic vein

Cephalic vein

Figure 6.17   The cephalic and accessory cephalic veins.

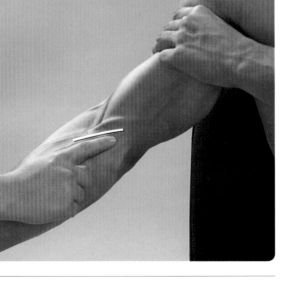

Figure 6.16   Locating the basilic vein.

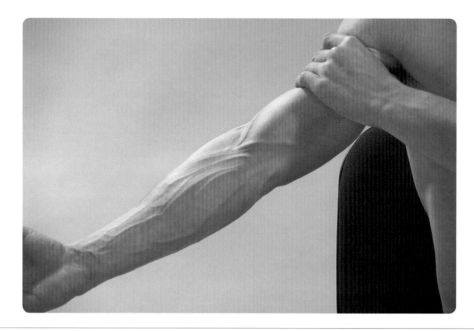

Figure 6.18   Venous network of the median vein of the forearm.

REVIEW OF TERMINOLOGY

The following terms were discussed in this chapter. Define or describe each term and, where appropriate, identify the location of the named structure either on your body or in an appropriate illustration.

accessory cephalic vein
arteries
arterioles
axillary artery
axillary nerve
axillary vein
basilic cubital vein
basilic vein
brachial artery
brachial plexus
brachial vein
brachiocephalic vein
capillaries
cephalic vein
deep veins
deep volar arch
digital arteries
digital veins
dorsal carpal branch
dorsal radial carpal branch
dorsal scapular nerve
dorsal thoracic nerve

infraclavicular nerves
internal jugular vein
lateral anterior thoracic nerve
lateral aspect of the median nerve
lateral cord
latissimus dorsi
long thoracic nerve
lower subscapular nerve
medial anterior thoracic nerve
medial aspect of the median nerve
medial brachial cutaneous nerve
medial cord
medial cutaneous nerve of the forearm
median cubital vein
median nerve
median vein
musculocutaneous nerve
musculospiral nerve
nerves
palmar venous vessels
plexus

posterior cord
radial artery
radial nerve
radial vein
smooth muscle fibers
subclavian artery
subclavian nerve
subclavian vein
subscapular nerve
superficial veins
superficial volar arch
supraclavicular nerves
suprascapular nerve
thoracodorsalis
ulnar artery
ulnar nerve
ulnar vein
upper subscapular nerve
veins
venous arch
venules

SUGGESTED LEARNING ACTIVITIES

1. With your hand in front of your body and your fingers facing upward, make a fist four to six times by opening and closing your hand. After your last effort, keep your hand closed in a fist. Apply pressure with your other hand, with your index finger and thumb to either side of your wrist (on the volar surface), as hard as you can. Continue the pressure while opening your hand (releasing the fist formation). Note the color of the palmar surface of your hand.

   a. Release the pressure being applied by your thumb. What happened to the color of your hand? Which side of your hand changed first? What vessels were compressed and released?

   b. Release the pressure being applied by your index finger. What happened to the color of your hand? Which side of your hand changed first? What vessels were compressed and released?

2. Have your partner encircle your upper arm just below the axilla with his or her hands. As your partner squeezes your arm, flex your elbow and wrist isometrically for a few seconds.

   a. What, if any, superficial veins of the arm and forearm can you identify?

   b. Trading positions with your partner, perform the same maneuver. Were the same veins identifiable, and were they in the same location?

MULTIPLE-CHOICE QUESTIONS

1. The ulnar and radial arteries are branches of which of the following arteries?

   a. dorsal carpal
   b. brachial
   c. volar carpal
   d. median

2. How many cords of nerves make up the brachial plexus?

   a. one
   b. two
   c. three
   d. four

3. The waste material produced by the production of energy within muscle tissue is disposed of through the

   a. arteries
   b. neutralizers
   c. veins
   d. kidneys

4. Which of the following veins drains directly into the subclavian vein?

   a. axillary
   b. ulnar
   c. brachial
   d. radial

FILL-IN-THE-BLANK QUESTIONS

1. When someone either donates blood or has blood drawn for testing purposes, the _____ vein is likely to be the vein the blood is drawn from.

2. The levator scapulae and rhomboid muscles are innervated by the _____ nerve.

3. The walls of the blood vessels known as _____ _____ are lined with smooth muscle fibers.

PART II SUMMARY TABLE Articulations of the Upper Extremity

Joint	Type	Bones	Ligaments	Movement
Shoulder girdle				
Sternoclavicular	Arthrodial (saddle/gliding)	Sternum and clavicle	• Sternoclavicular (anterior, superior, and posterior) • Costoclavicular • Interclavicular • (Articular disc)	Elevation/depression, rotation, protraction/retraction
Acromioclavicular	Arthrodial (plane)	Clavicle and scapula	• Acromioclavicular • Coracoclavicular (conoid and trapezoid) • Coracoacromial (for protection of glenohumeral joint) • (Articular disc)	Elevation/depression, rotation, protraction/retraction, winging (posterior movement to keep scapula close to the thorax during the abduction of the scapula), tipping (rotation to keep scapula close to the thorax during elevation of scapula)
Shoulder joint				
Glenohumeral	Ball and socket	Humerus and scapula	• Glenohumeral (superior, middle, and inferior) • Coracohumeral • Transverse humeral (keeps long-head biceps tendon in intertubercular groove) • Labrium (deepens socket of the glenoid fossa) • Capsule	Flexion, extension, (hyperextension), abduction, adduction, internal and external rotation
Scapulothoracic	(False joint)	Scapula and thorax	• No ligaments (fascia of the serratus anterior muscle and fascia from the thorax)	Elevation, depression, abduction, (protraction), adduction, (retraction), upward and downward rotation and tilt of the scapula on the thorax
Elbow				
Humeroulnar	Hinge	Humerus and ulna	• Capsule • Medial (ulnar) collateral (3 bands: anterior, posterior, and transverse)	Flexion and extension
Radiohumeral	Irregular (gliding)	Humerus and radius	• Lateral (radial) collateral • Annular	Flexion, extension, and rotation (pronation/supination)
Radioulnar (proximal)	Pivot	Radius and ulna	• Annular	Pronation and supination
Forearm				
Radioulnar (middle)	Syndesmosis	Radius and ulna	• Oblique cord • Interosseous membrane	Slight (if any)
Radioulnar (distal)	Pivot	Radius, ulna (and articular disc)	• Capsule • Dorsal radioulnar • Volar (palmar) radioulnar	Pronation and supination
Wrist				
Radiocarpal	Condyloid (gliding)	Radius, scaphoid, lunate, triquetrum	• Capsule • Radial collateral • Ulnar collateral • Dorsal radioulnar • Volar (palmar) radioulnar	Flexion, extension, abduction, (radial deviation), adduction, (ulnar deviation)

Wrist				
Ulnarcarpal (not part of the wrist)	Irregular (gliding)	Ulna, pisiform, triquetrum	• Capsule • Ulnar collateral • Articular disc	Gliding
Wrist: intercarpals				
Proximal row	Arthrodial (gliding)	Radius, scaph-oid, lunate, triquetrum	• Capsule • Flexor (volar) retinaculum • Extensor (dorsal) retinaculum • Radial collateral • Ulnar collateral • Volar (palmar) intercarpal • Dorsal intercarpal	Gliding
Middle	Hinge (gliding)	Proximal and distal rows of carpals	• Capsule • Volar (palmar) intercarpal • Dorsal intercarpal • Interosseous intercarpal • Radial collateral • Ulnar collateral • Pisohamate • Pisometacarpal	Gliding
Distal row	Irregular (gliding)	Trapezius, trap-ezoid, capitate, hamate	• Capsule • Volar (palmar) intercarpal • Dorsal intercarpal • Interosseous intercarpal	Gliding
Hand and fingers				
Carpometacarpal	Modified saddle	Trapezium, trap-ezoid, capitate, hamate, (4) metacarpals	• Capsule • Volar carpometacarpal • Dorsal carpometacarpal • Interosseous carpometacarpal • (Volar accessory)	Flexion, extension, abduc-tion, adduction, slight opposition of fifth
Metacarpophalangeal	Condyloid	(4) metacarpals, (4) proximal phalanges	• Transverse metacarpal • Capsule • Radial collateral • Ulnar collateral • (Volar accessory)	Flexion, extension, abduc-tion, adduction,
Proximal interphalan-geal	Hinge	(4) proximal phalanges and (4) middle phalanges	• Capsule • Radial collateral • Ulnar collateral • (Volar accessory)	Flexion, extension
Distal interphalangeal	Hinge	(4) middle phalanges and (4) distal phalanges	• Capsule • Radial collateral • Ulnar collateral • (Volar accessory)	Flexion, extension
Thumb				
Carpometacarpal	Saddle	Trapezium and 1st metacarpal	• Capsule	Flexion, extension, abduc-tion, adduction, opposition
Metacarpophalangeal	Hinge	1st metacarpal and proximal phalanx	• Capsule • Radial collateral • Ulnar collateral • (Dorsal accessory)	Flexion, extension
Interphalangeal	Hinge	Proximal phalanx and distal pha-lanx	• Capsule • Radial collateral • Ulnar collateral • (Volar accessory)	Flexion, extension

PART II   SUMMARY TABLE   Upper-Extremity Muscles, Nerves, and Blood Supply

Muscle	Origin	Insertion	Action	Nerve	Blood supply
Shoulder girdle, anterior					
Pectoralis minor	3rd to 5th ribs	Coracoid process of scapula	Downward rotation of scapula	Medial anterior thoracic	Lateral thoracic and thoraco-acromial branches of axillary
Serratus anterior	Upper 9 ribs	Vertebral border of scapula	Upward rotation of scapula	Long thoracic	Lateral thoracic and subscapular branches of axillary, transverse cervical branch of thyrocervical branch of subclavian
Subclavian	1st rib	Subclavian groove of clavicle	Ligamentous action at sternoclavicular joint	Medial anterior thoracic	Clavicular branch of thoracoacromial branch of axillary thoracic
Shoulder girdle, posterior					
Levator scapulae	Transverse processes of first 4 cervical vertebrae	Vertical border of scapula	Elevation and downward rotation of scapula	Dorsal scapular	Transverse cervical branch of thyrocervical and intercostals
Rhomboids (major and minor)	Spines of vertebrae C7 to T5	Vertebral border of scapula	Elevation and downward rotation of scapula	Dorsal scapular	Transverse cervical branch of thyrocervical branch of the subclavian
Trapezius	External occipital protuberance, all 7 cervical and 12 thoracic spinous processes	Spine of scapula, acromion process, lateral 1/3 of posterior clavicle	Elevation, upward and downward rotation, and adduction of scapula	C2, C3, C4, and spinal accessory (11th cranial)	Transverse cervical branch of thyrocervical branch of the subclavian
Shoulder joint, anterior					
Pectoralis major	2nd to 6th ribs, sternum, medial half of clavicle	Inferior to greater tuberosity of humerus in area of surgical neck	Flexion, adduction, and internal rotation of GH joint	Lateral and medial anterior thoracic	Lateral thoracic and thoracoacromial branches of axillary
Coracobrachialis	Coracoid process of scapula	Middle 1/3 of medial surface of humerus	Flexion and adduction of GH joint	Musculocutaneous	Brachial

Muscle	Origin	Insertion	Action	Nerve	Artery
Biceps brachii (long head)	Supraglenoid tubercle	Radial tuberosity	Abduction and flexion of GH joint, flexion of elbow, supination of forearm	Musculocutaneous	Brachial
Biceps brachii (short head)	Coracoid process of scapula	Radial tuberosity	Adduction and flexion of GH joint, flexion of elbow, supination of forearm	Musculocutaneous	Brachial
Subscapularis	Subscapular fossa	Lesser tuberosity of humerus	Internal rotation and flexion of GH joint, GH joint stability	Subscapular branch from brachial plexus	Subscapular branch of axillary, transverse cervical and transverse scapular branches of thyrocervical
Shoulder joint, superior					
Deltoid (clavicular)	Lateral 1/3 of anterior border of clavicle	Deltoid tubercle	Flexion, adduction, and internal rotation of GH joint (above horizontal, abduction)	Axillary	Anterior and posterior humeral circumflex, brachial, thoraco-acromial
Deltoid (acromial)	Lateral border of acromion process	Deltoid tubercle	Abduction of GH joint		
Deltoid (scapular)	Inferior lip of spine of scapula	Deltoid tubercle	Extension, adduction, and external rotation of GH joint (above horizontal, abduction)		
Supraspinatus	Supraspinous fossa of scapula	Proximal facet of greater tuberosity of humerus	Abduction of GH joint, GH joint stability	Suprascapular from C5 through brachial plexus	Transverse cervical and transverse scapular branches of thyrocervical and subscapular
Shoulder joint, posterior					
Infraspinatus	Infraspinatus fossa of spine of scapula	Middle facet of greater tuberosity of humerus	External rotation and extension of GH joint, GH joint stability	Suprascapular through brachial plexus	Transverse cervical and transverse scapular branches of thyrocervical
Teres minor	Upper 2/3 of lateral border of scapula	Distal facet of greater tuberosity of humerus	External rotation, extension, and adduction of GH joint, GH joint stability	Axillary through brachial plexus	Posterior humeral circumflex and subscapular

(continued)

Muscle	Origin	Insertion	Action	Nerve	Blood supply
Shoulder joint, inferior					
Latissimus dorsi	Spinous processes of lower 6 thoracic and all lumbar vertebrae, ilium, lower 3 ribs, inferior angle of scapula	Intertubercular groove	Internal rotation, extension, and adduction of GH joint	Thoracodorsal nerve and posterior rami of lower 6 thoracic and lumbar spinal nerves	Subscapular branch of axillary, transverse cervical branch of thyrocervical branch of subclavian
Teres major	Inferior 1/3 of lateral border and inferior angle of scapula	Beneath the lesser tuberosity on anterior humerus	Internal rotation, extension, and adduction of GH joint	Subscapular through brachial plexus	Subscapular branch of axillary
Triceps brachii (long head)	Infraglenoid tubercle of scapula	Olecranon process of ulna	Extension and adduction of GH joint	Radial	Posterior humeral circumflex, profundus branch of brachial
Elbow joint, anterior					
Brachialis	Distal 1/3 of anterior surface of humerus	Coronoid process and tuberosity of ulna	Flexion of elbow	Musculocutaneous	Brachial
Brachioradialis	Proximal aspect of lateral supracondylar ridge of humerus	Lateral aspect of styloid process of radius	Flexion of elbow, supination of forearm	Radial	Radial
Biceps brachii (long head)	Supraglenoid tubercle	Radial tuberosity	Abduction and flexion of GH joint, flexion of elbow, supination of forearm	Musculocutaneous	Brachial
Biceps brachii (short head)	Coracoid process of scapula	Radial tuberosity	Adduction and flexion of GH joint, flexion of elbow, supination of forearm	Musculocutaneous	Brachial
Elbow joint, posterior					
Triceps brachii (lateral head)	Proximal 1/3 of posterolateral aspect of humerus	Olecranon process of ulna	Extension of elbow	Radial	Posterior humeral circumflex, profundus branch of brachial
Triceps brachii (long head)	Infraglenoid tubercle of scapula	Olecranon process of ulna	Extension of the elbow, extension and adduction of GH joint	Radial	Posterior humeral circumflex, profundus branch of brachial
Triceps brachii (medial head)	Distal 2/3 of posteromedial aspect of humerus	Olecranon process of ulna	Extension of elbow	Radial	Posterior humeral circumflex, profundus branch of brachial

Muscle	Origin	Insertion	Action	Nerve	Blood supply
Anconeus	Lateral epicondyle of humerus	Radial aspect of olecranon process and dorsal surface of proximal 1/4 of ulna	Extension of elbow	Radial	Profundus branch of brachial and dorsal interosseous

Forearm, supinators

Muscle	Origin	Insertion	Action	Nerve	Blood supply
Supinator	Lateral epicondyle of humerus and supinator fossa of ulna	Proximal 1/3 of lateral and volar aspect of radius	Supination of forearm	Radial	Dorsal interosseous and radial
Biceps brachii (long head)	Supraglenoid tubercle	Radial tuberosity	Abduction and flexion of GH joint, flexion of elbow, supination of forearm	Musculocutaneous	Brachial
Biceps brachii (short head)	Coracoid process of scapula	Radial tuberosity	Adduction and flexion of GH joint, flexion of elbow, supination of forearm	Musculocutaneous	Brachial

Forearm, pronators

Muscle	Origin	Insertion	Action	Nerve	Blood supply
Pronator teres	Common flexor tendon and coronoid process of ulna	Middle of lateral aspect of radius	Pronation of forearm	Median	Radial branch of brachial
Pronator quadratus	Distal part of volar aspect of ulna	Distal part of volar aspect of radius	Pronation of forearm	Median	Ulnar and radial

Wrist and hand, anterior wrist and extrinsic hand

Muscle	Origin	Insertion	Action	Nerve	Blood supply
Flexor carpi radialis	Common flexor tendon	Bases of 2nd and 3rd metacarpals	Flexion and radial deviation of wrist, elbow flexion, and forearm pronation	Median	Radial
Flexor carpi ulnaris	Common flexor tendon	Pisiform, hamate, and base of 5th metacarpal	Flexion and ulnar deviation of wrist, elbow flexion, and forearm supination	Ulnar	Ulnar branch of brachial
Flexor digitorum superficialis	Common flexor tendon, coronoid process, and proximal 2/3 of radius	Split tendons on each side of base of middle phalanges of the 4 fingers	Flexion of PIP, MP, and wrist joints	Median	Ulnar branch of brachial
Palmaris longus	Common flexor tendon	Palmar aponeurosis	Flexion of wrist	Median	Radial branch of brachial
Flexor digitorum profundus	Proximal 3/4 of coronoid process of ulna and interosseous membrane	4 tendons—passing split superficialis tendons—on bases of distal phalanges of the 4 fingers	Flexion of DIP, PIP, MP, and wrist joints	Median and ulnar	Ulnar and volar interosseous

(continued)

PART II SUMMARY TABLE *(continued)*

Muscle	Origin	Insertion	Action	Nerve	Blood supply
Wrist and hand, posterior wrist and extrinsic hand					
Extensor carpi radialis longus	Lateral supracondylar ridge of humerus	Base of 2nd metacarpal	Extension and radial deviation of wrist, supination of forearm	Radial	Radial
Extensor carpi radialis brevis	Common extensor tendon	Base of 3rd metacarpal	Extension and radial deviation of wrist, supination of forearm	Radial	Radial
Extensor carpi ulnaris	Common extensor tendon	Base of 5th metacarpal	Extension and ulnar deviation of wrist	Radial	Ulnar
Extensor digitorum communis	Common extensor tendon	Bases of distal phalanges of 4 fingers	Extension of DIP, PIP, and MP joints of 4 fingers, extension of wrist	Radial	Ulnar
Extensor digiti minimi (proprius)	Common extensor tendon	Base of proximal phalanx of little finger	Extension of 5th MP joint and wrist	Radial	Ulnar
Extensor indicis	Dorsal aspect of distal ulna and interosseous membrane	Base of proximal phalanx of index finger	Extension and radial deviation of 2nd MP joint, extension of wrist	Radial	Ulnar
Wrist and hand, intrinsic hand					
Abductor digiti minimi	Pisiform bone	Base of proximal phalanx of little finger	Abduction and flexion of 5th MP joint	Ulnar	Ulnar
Flexor digiti minimi brevis	Hamate bone	Base of proximal phalanx of little finger	Flexion of 5th MP joint	Ulnar	Ulnar
Opponens digiti minimi	Hamate bone	Medial surface of 5th metacarpal	Flexion and adduction of little finger	Ulnar	Ulnar
Dorsal interossei (4)	Adjacent sides of all 4 metacarpals	Bases of 2nd, 3rd, and 4th proximal phalanges and extensor digitorum communis tendon	Abduction of 2nd and 4th MP joints	Ulnar	Metacarpal branches of radial

Muscle	Origin	Insertion	Action	Innervation	Blood supply
Palmar interossei (3)	Ulnar side of 2nd metacarpal and radial side of 4th and 5th metacarpals	Radial side of 4th and 5th phalanges and ulnar side of 2nd proximal phalanx	Adduction and flexion of 2nd, 4th, and 5th MP joints	Ulnar	Metacarpal branches of deep volar arch
Lumbricales	Flexor digitorum profundus tendons	Extensor digitorum communis tendons	Flexion of 4 MP joints, extension of 4 PIP and DIP joints	Median and ulnar	Volar metacarpal branch of deep volar
Thumb, extrinsic					
Extensor pollicis longus	Middle 1/3 of ulna and interosseous membrane	Base of distal phalanx of thumb	Extension of IP and MP joints of thumb, extension and radial deviation of wrist	Radial	Ulnar through the dorsal interosseous
Extensor pollicis brevis	Middle dorsal aspect of radius and interosseous membrane	Base of proximal phalanx of thumb	Extension of MP joint of thumb, abduction of 1st metacarpal, radial deviation of wrist	Radial	Ulnar through the dorsal interosseous
Abductor pollicis longus	Middle dorsolateral aspect of ulna and interosseous membrane	Base of 1st metacarpal	Abduction of 1st metacarpal, flexion of wrist	Radial	Dorsal interosseous branch of ulnar
Flexor pollicis longus	Middle 1/2 of volar surface of radius	Base of distal phalanx of thumb	Flexion of IP and MP joints, adduction of thumb	Median	Radial, volar interosseous
Thumb, intrinsic					
Abductor pollicis brevis	Trapezius and scaphoid bones	Base of proximal phalanx of thumb	Abduction of MP joint of thumb	Median	Radial
Flexor pollicis brevis	Trapezium	Base of proximal phalanx of thumb	Flexion and adduction of MP joint of thumb	Median	Radial
Opponens pollicis	Trapezium	Lateral surface of 1st metacarpal	Flexion and adduction of MP joint of thumb	Median	Radial
Adductor pollicis	Oblique head: capitate and 2nd and 3rd metacarpal bones; transverse head: 3rd metacarpal	Volar surface of base of proximal phalanx of thumb	Adduction of MP joint of thumb	Median	Deep volar arch

GH = glenohumeral; PIP = proximal interphalangeal; MP = metacarpophalangeal; DIP = distal interphalangeal; IP = interphalangeal.

The Spinal Column, Pelvis, and Thorax

The Spinal Column and Pelvis

The spinal column (figure 7.1) is a stack of 33 bones called vertebrae held together by ligaments and muscles, with cartilaginous discs (primarily water and protein) between the bones. All the vertebrae have many common characteristics, but each group has unique features designed for specific purposes. The 33 vertebrae are divided into five distinct sections. The most superior group is known as the cervical (neck) spine and contains seven vertebrae. The next group is known as the thoracic (chest) spine and contains 12 vertebrae. The next group is known as the lumbar (low back)

spine and contains five vertebrae. The next group is known as the sacral spine and contains five vertebrae fused together into one structure known as the sacrum. The last or most distal group is known as the coccygeal spine and contains four vertebrae fused together into one structure known as the coccyx.

The lateral view of the spinal column reveals four curvatures: anterior (convex) curves in the cervical and lumbar spine and posterior (concave) curves in the thoracic and sacral-coccygeal spine. These curvatures may increase

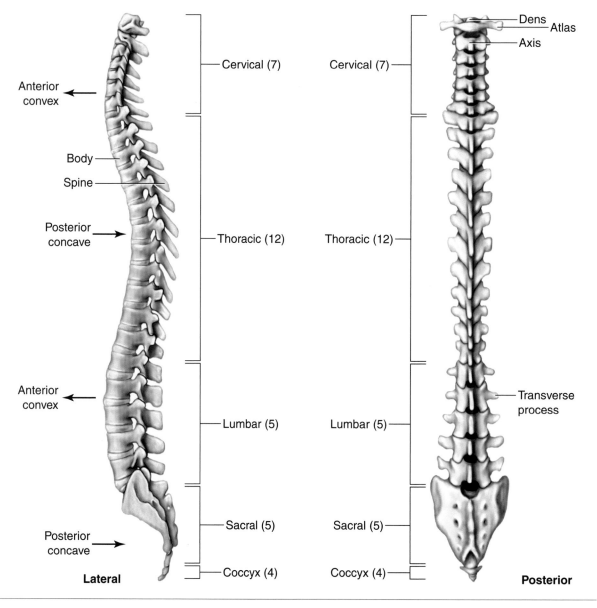

Figure 7.1   The spinal column, lateral and posterior views.

or decrease as the body's center of gravity shifts (e.g., with pregnancy, weight gain, weight loss, or trauma). This is a result of one of the spinal column's functions: to maintain, in the upright body position, the brain over the body's center of gravity. Over- or underdevelopment of the musculature on either side of the spinal column, structural deformities, or other causes can result in excessive curvatures of the spinal column. Three of the more common conditions resulting from excessive curvature are **kyphosis** (figure 7.2), excessive posterior curvature of the thoracic spine (hunchback, round shoulders); **lordosis** (figure 7.3), excessive anterior curvature of the lumbar spine (swayback); and **scoliosis** (figure 7.4), excessive lateral curvature of the spinal column, usually in the thoracic spine but sometimes to a lesser extent in the cervical and lumbar spine.

Figure 7.3 Lumbar lordosis.

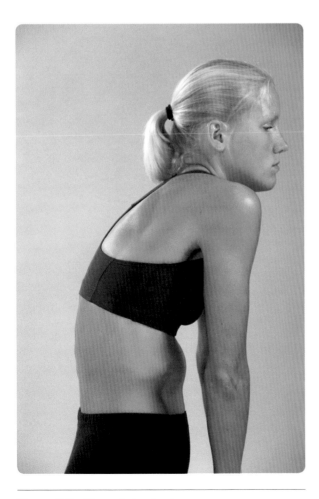

Figure 7.2 Thoracic kyphosis.

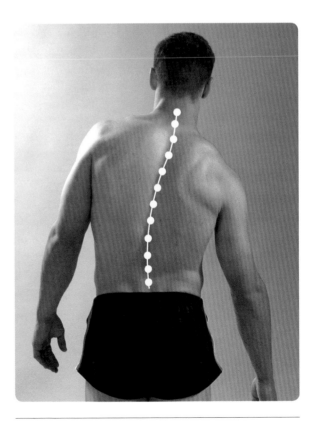

Figure 7.4 Scoliosis.

Bones of the Spinal Column

Although all five sections of the spinal column have common structures, the vertebrae of each are spaced uniquely and may be shaped slightly differently, depending on their functions in a particular anatomical area. All vertebrae have a body, two transverse processes laterally (serving as sources for ligamentous and muscular attachments), a spinous process (serving as another source for ligamentous and muscular attachments), and a vertebral foramen (where the spinal cord and nerve roots pass) (figure 7.5). Additionally, each vertebra has the following features. Superior and inferior articulating facets are where a vertebra articulates with the vertebrae above and below it. A lamina forms the posterior aspect of the vertebral foramen. Pedicles form the lateral sides of the vertebral foramen. The intervertebral foramen between the vertebrae allows the nerve branches from the spinal cord to pass through. The isthmus (also called pars interarticularis or neck) is the bony area between the superior and inferior articulating facets.

The seven **cervical vertebrae** are numbered from the most superior to most inferior as C1, C2, and so on. The first cervical vertebra, the **atlas** (C1), and the second cervical vertebra, the **axis** (C2), are shaped differently from the other five cervical vertebrae (C3–C7) to permit the head to rotate. The atlas has no significant body but has two large articular facets that provide the surface where the skull and the spinal column articulate (figure 7.6). The atlas slides over the axis and rests on top of the two large superior articulating surfaces of the axis, between which lies a rather large bony process from the axis body known as the **dens,** or **odontoid process** (figure 7.6). Note also that the cervical vertebrae have a **bifid** (split) spinous process and a foramen in each transverse process to provide for the passage of blood vessels through the cervical spine (figure 7.7). These two features are unique to the cervical vertebra. One other difference to note is the rather long and prominent spinous process of the seventh cervical vertebra (C7). This prominence is easily palpated.

Hands on . . . Run your finger down the posterior aspect of your cervical spine until you find a large protrusion. This is the C7 spinous process, which serves as an anatomical landmark for determining the spinous processes of the other cervical vertebrae and determining where the thoracic spine begins.

The 12 **thoracic vertebrae** have bony features similar to all other vertebrae, with a few unique features. Note the longer and more vertical spinous

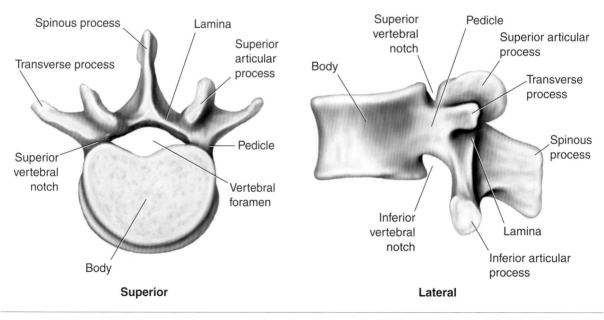

Superior   **Lateral**

Figure 7.5 A typical vertebra, superior view and lateral view.

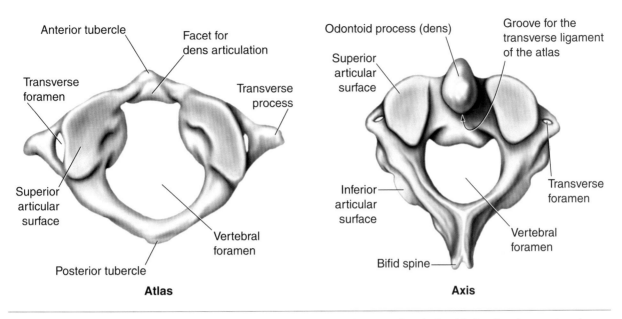

Figure 7.6   The first cervical vertebra (atlas) and the second cervical vertebra (axis with the dens), superior view.

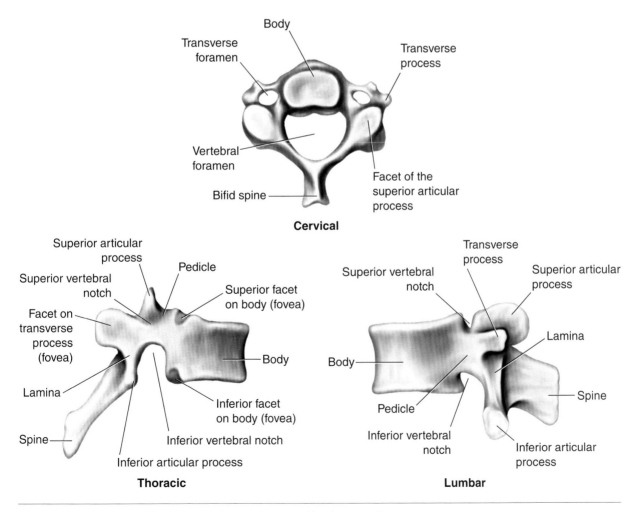

Figure 7.7   Cervical vertebrae, thoracic vertebrae, and lumbar vertebrae.

processes (figure 7.7). Also note the articulating surfaces (fovea) on the anterior lateral aspects of the transverse processes and on the superior and inferior portions of the posterior lateral aspects of the vertebral bodies. These notches provide the articulation of the 12 pairs of ribs with the 12 thoracic vertebrae.

The five **lumbar vertebrae** are the largest vertebrae (figure 7.7). They have no foramen through their transverse processes, nor are there any articular facets (fovea) on their bodies. Although separate at birth, the five **sacral vertebrae** (S1–S5) fuse together to form a large triangular-shaped bone known as the **sacrum** (figures 7.8 and 7.9) during the growth process. The two large articular surfaces formed on the lateral aspects of the sacrum are where the spinal column articulates with the bones of the pelvis, forming the pelvic girdle.

The **coccyx** (the final four vertebrae), like the sacrum, is originally four vertebrae that, during the growth process, fuse to form one structure (figure 7.10). It serves as a source of attachment for ligamentous and muscular structures. Some individuals refer to this structure as humans' "vestigial tail," in reference to the evolution of humans from species that had tails.

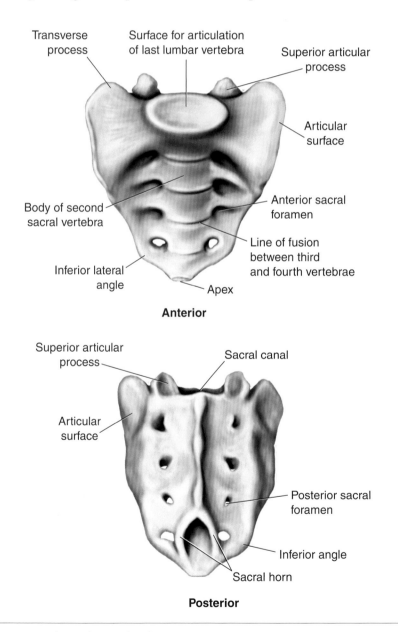

Figure 7.8   The sacrum, anterior and posterior views.

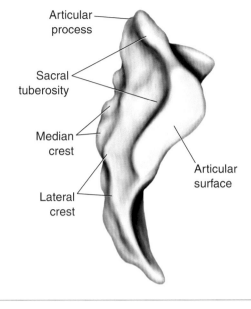

Figure 7.9   The sacrum, lateral view.

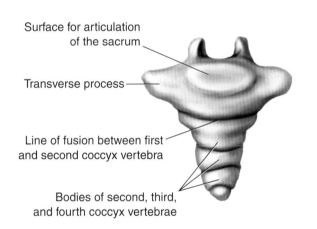

Figure 7.10   An anterior view of the coccyx.

Ligaments of the Spinal Column

There is only one movable joint in the skull, the temporomandibular joint, which enables the opening and closing of the mouth; any movement of the head is the result of movement at the joints between the occipital bone of the skull and the first and second cervical vertebrae. There are approximately 25 such ligaments (figures 7.11 and 7.12). With the exception of one ligament discussed later, we will not devote time to the individual ligaments but simply note the groups they belong

to. **Atlanto-occipital ligaments** attach the occipital bone of the skull to the first cervical vertebra (atlas, or C1). **Occipitoaxial ligaments** attach the occipital bone of the skull to the dens (odontoid process) of the axis, and **atlantoaxial ligaments** attach the atlas and the axis. A final group of ligaments is not involved in movement of the spinal column: the costovertebral ligaments (six per rib) articulate the 12 pairs of ribs of the thorax with the 12 thoracic vertebrae.

One of the atlantoaxial ligaments, the **transverse ligament** (figure 7.13), runs from one

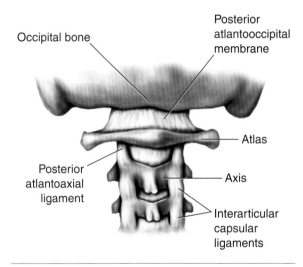

Figure 7.11   Atlanto-occipital and atlantoaxial ligaments of the cervical spine.

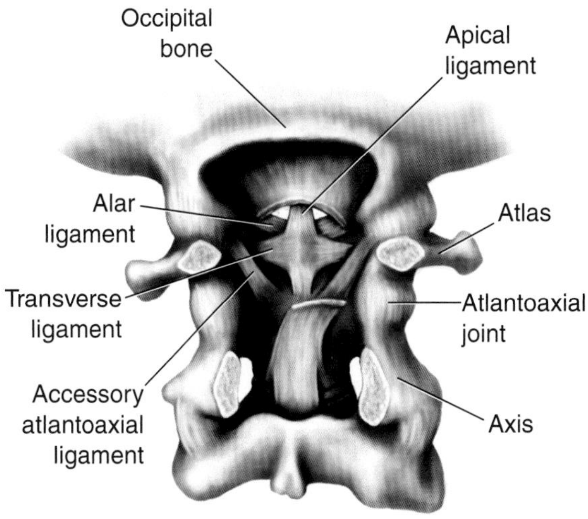

Figure 7.12   Occipitoaxial and atlantoaxial ligaments of the cervical spine.

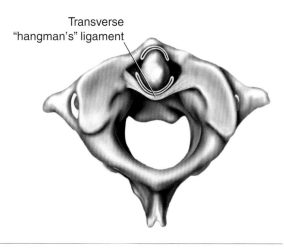

Figure 7.13   The transverse ligament ("hangman's" ligament) of the atlas (C1).

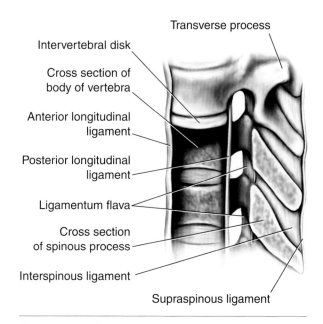

Figure 7.14   The ligaments of the spinal column.

transverse process of the atlas (C1), across the vertebral foramen, to the other transverse process and holds the dens (odontoid process) of the axis (C2) in place. Failure of this ligament to restrain movement of the dens of C2 beneath C1 could cause the dens to do irreparable damage to structures in the vertebral foramen at the C1–C2 level of the spinal column. Many authors nickname the transverse ligament the "hangman's" ligament, because the skull and C1 of a person being hanged are restricted from downward movement by a hangman's noose, while the rest of the body from C2 and below moves downward through the pull of gravity. The dens of C2 tears the transverse ligament of C1 and compresses the spinal cord against C1, disrupting nerves to the heart and lungs.

From the most superior to the most inferior aspects of the spinal column are a number of ligaments that play a major role in the movement of the joints between the vertebrae (figure 7.14). Two ligaments known as the **interbody ligaments** run the entire length of the spinal column. The **anterior longitudinal ligament** runs along the anterior aspect of the bodies of all 33 vertebrae. This ligament is structurally the weakest of all the spinal column ligaments. The **posterior longitudinal ligament** runs along the posterior aspect of the bodies of all 33 vertebrae. The posterior longitudinal ligament forms the anterior wall of the spinal canal.

The **ligamentum flavum** runs between the laminae of successive vertebrae (figure 7.14).

The **interspinous ligament** runs between successive vertebrae's spinous processes. Running between the dorsal tips of each vertebra's spinous process from the coccyx to the external

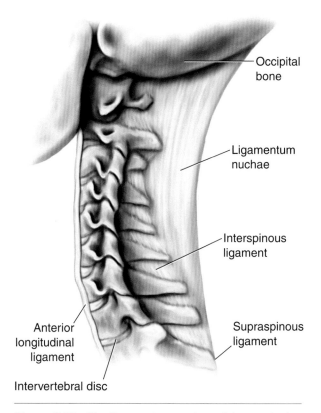

Figure 7.15   The ligamentum nuchae of the cervical spine.

occipital protuberance of the occipital bone is the **supraspinous ligament**. Between the external occipital protuberance and the spinous process of the seventh cervical vertebra, the supraspinous ligament is known as the **ligamentum nuchae** (figure 7.15).

One additional ligament of the spinal column is the iliolumbar ligament, which runs between the transverse processes of the fifth lumbar vertebra to the ilium of the pelvis (figure 7.16). The joints and ligaments between the spinal column and the pelvis are presented later in this chapter.

One other structure of the spinal column is the **intervertebral disc** (figure 7.17). These cartilaginous (primarily water and protein) discs lie on the bodies of each vertebra and serve both as spacers (to help separate the vertebrae and allow nerve roots to pass from the spinal canal to other structures of the body) and as shock absorbers for the spinal column. The disc has two distinct portions: The inner portion is known as the **nucleus pulposus,** and the outer portion is known as the

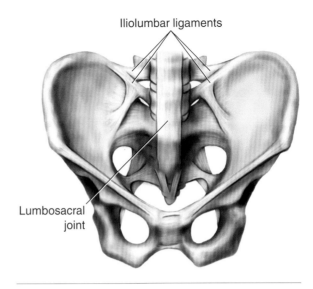

Iliolumbar ligaments

Lumbosacral joint

Figure 7.16　The lumbosacral joint and iliolumbar ligament, anterior view.

annulus fibrosus. The annulus fibrosus consists of fibrous tissue, whereas the nucleus pulposus consists of soft, pulpy, elastic tissue.

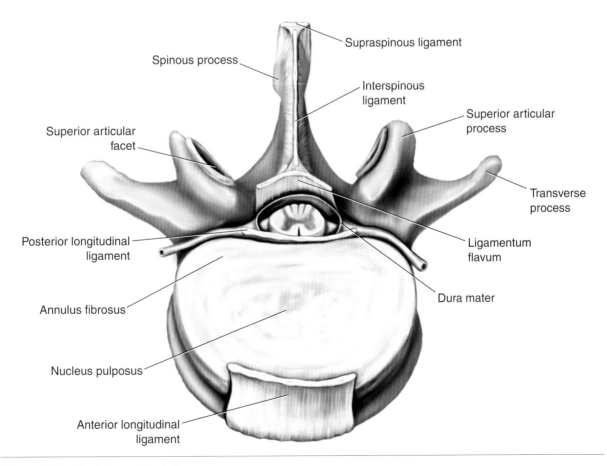

Spinous process

Superior articular facet

Superior articular process

Supraspinous ligament

Interspinous ligament

Transverse process

Posterior longitudinal ligament

Ligamentum flavum

Annulus fibrosus

Dura mater

Nucleus pulposus

Anterior longitudinal ligament

Figure 7.17　The intervertebral disc.

Focus on . . . THE INTERVERTEBRAL DISC

The terms *slipped disc* and *herniated disc* are frequently heard in reference to someone having back pain. If other complications are ruled out (e.g., muscle strain, ligamentous sprain), a slipped disc means the intervertebral disc has protruded into the space occupied by either the spinal cord or its nerve roots, compressing the nerve and causing pain and disability. A more technical term would be *protruded disc.* The term *herniated disc* refers to the actual tearing of the disc structure that allows a portion of the disc material to escape and cause pressure on the spinal cord or its nerve roots. Both the protruded disc and herniated disc can occur as a result of excessive motion of the spinal column (hyperflexion, hyperextension, lateral flexion, or rotation).

A herniated disc in the lumbar spine can be detected from a spinal tap, or lumbar puncture, a procedure in which fluid is removed from the space beneath the arachnoid membrane (which surrounds the spinal cord) of the lumbar region of the spinal cord. Elevated levels of a telltale protein indicate a herniated disc. Disc problems need to be treated by neurosurgeons and orthopedic surgeons.

Movements and Muscles of the Spinal Column

Movements of the joints between the vertebrae of the spinal column occur in all three planes about all three axes. The fundamental movements of the spinal column (the cumulative actions of all the joints among the 33 vertebrae) are flexion, extension, lateral flexion (as opposed to abduction and adduction in the extremities), and rotation (figures 7.18 and 7.19). The greatest amount of movement of the spinal column takes place in the cervical and lumbar regions. Movement is more restricted

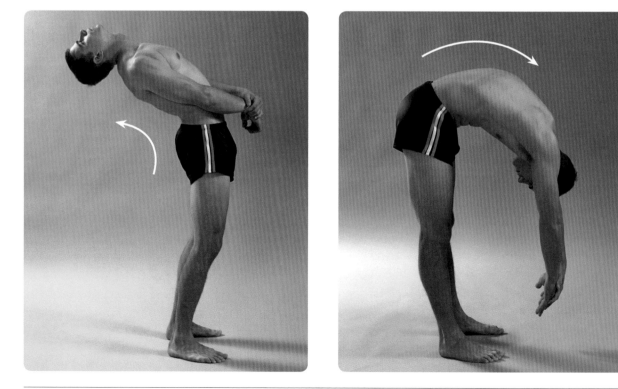

Figure 7.18   Extension and flexion of the spine.

Figure 7.19   Rotation and lateral flexion of the spine.

in the thoracic region because of the attached ribs. There is no movement in the sacral and coccygeal regions because the vertebrae in these regions are fused together into the five-bone sacrum and the four-bone coccyx.

The muscles responsible for the movements of the spinal column are, for the most part, either anterior or posterior to the spinal column. Those anterior to the spinal column are muscles that flex, laterally flex, or rotate the spine. Those posterior to the spinal column are muscles that extend, laterally flex, or rotate the spine. Some spinal column muscles are specific to a particular area of the spinal column (cervical, lumbar), whereas others have branches in multiple areas of the spinal column.

In the cervical region, the major anterior muscles and muscle groups are the **sternocleidomastoid**, the **prevertebrals (rectus capitis anterior, rectus capitis lateralis, longus capitis, longus colli)**, and the **scaleni (scalenus anterior, scalenus medius, scalenus posterior)** (figures 7.20 and 7.21). The sternocleidomastoid muscle,

as its name indicates, attaches to the sternum, the clavicle, and the mastoid process of the skull just posterior to the ear.

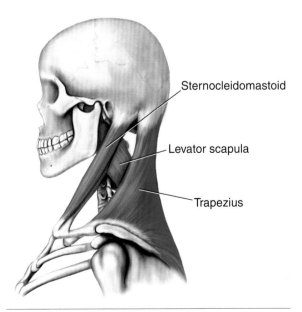

Figure 7.20   The sternocleidomastoid.

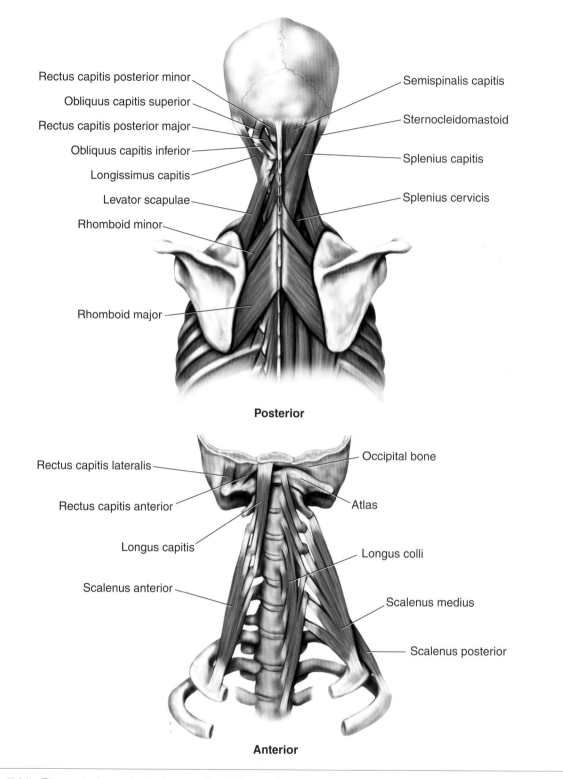

Rectus capitis posterior minor

Obliquus capitis superior

Rectus capitis posterior major

Obliquus capitis inferior

Longissimus capitis

Levator scapulae

Rhomboid minor

Rhomboid major

Semispinalis capitis

Sternocleidomastoid

Splenius capitis

Splenius cervicis

Posterior

Rectus capitis lateralis

Rectus capitis anterior

Longus capitis

Scalenus anterior

Occipital bone

Atlas

Longus colli

Scalenus medius

Scalenus posterior

Anterior

Figure 7.21   The posterior and anterior muscles of the cervical spine.

Hands on . . . Referring to figure 7.22, palpate the sternal and clavicular portions of the sternocleido-mastoid muscle. Use simple reasoning and realize that there is both a right and left sternocleido-

mastoid muscle: If this muscle lies anterior and slightly lateral to the cervical spine, what actions is it capable of performing? (The answer is flexion, lateral flexion, and rotation of the cervical spine.)

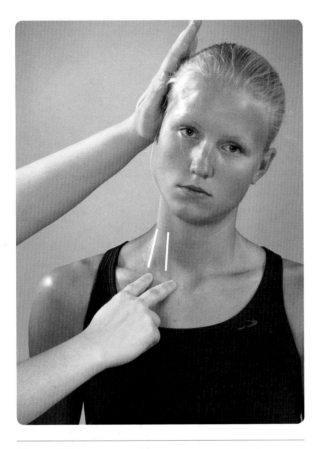

Figure 7.22 Locating the sternal and clavicular parts of the sternocleidomastoid.

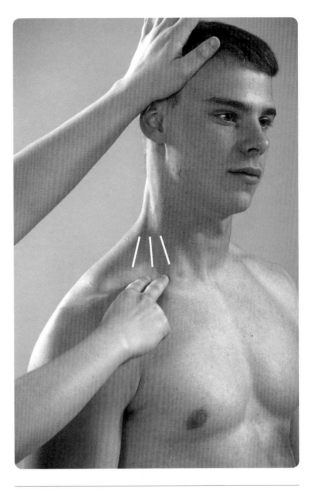

Figure 7.23 Identifying the scalenes.

The recti muscles (rectus capitis anterior, rectus capitis lateralis), the longus capitis, and the longus colli flex and rotate the cervical spine. The three scalene muscles (scalenus anterior, scalenus medius, scalenus posterior) run from the cervical vertebrae to the first three ribs. The scaleni not only laterally flex the cervical spine but, during forced respiration, also elevate the first three ribs to allow the lungs to expand.

Hands on . . . Isolating all three of the scalene muscles is difficult, but you should be able to locate the group by having your partner laterally flex his or her neck against your resistance (figure 7.23).

Restricted to the posterior aspect of the cervical spine are the rectus **(rectus capitis posterior major and minor),** obliquus **(obliquus capitis superior and inferior),** and splenius **(splenius capitis and cervicis)** muscle groups (see figure 7.21). These muscles are involved in the extension, lateral flexion, and rotation of the cervical spine.

Muscles of the shoulder girdle (presented in chapter 3) also are involved in cervical spine movement. The **levator scapulae,** the **trapezius,** and the most superior portion of the **rhomboids** all have attachments to the cervical spine and assist with the extension, lateral flexion, and rotation of the cervical spine (figure 7.24). Other muscles also are involved in cervical spine movement, but they are not specific to the cervical spine and are presented later in this chapter.

The muscles anterior to the thoracic and lumbar regions of the spinal column are often classified as the abdominal muscles: the **rectus abdominis,** the **obliquus internus** and **externus,** and the **transversus abdominis** (figure 7.25).

Hands on . . . Having your partner perform and hold an abdominal curl should reveal both the rectus abdominis and obliquus externus muscles (figures 7.26 and 7.27).

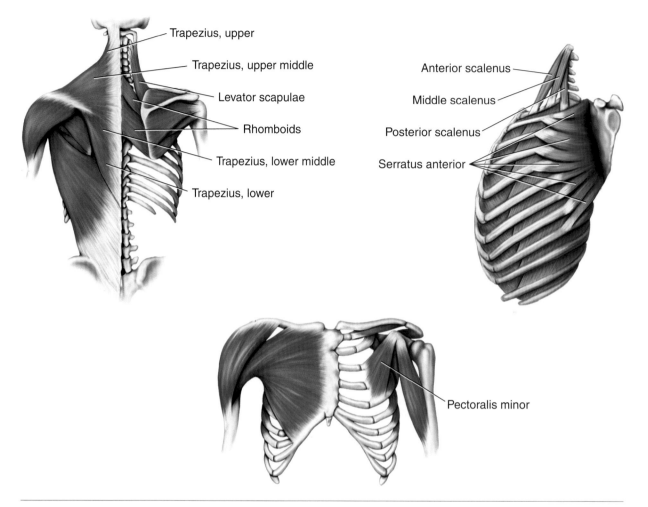

Figure 7.24   The superficial muscles acting at the scapula.

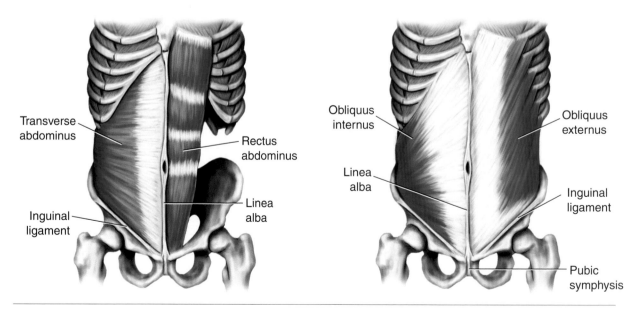

Figure 7.25   The linea alba, rectus abdominis, obliquus externus, obliquus internus, and transversus abdominis.

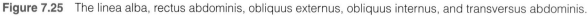

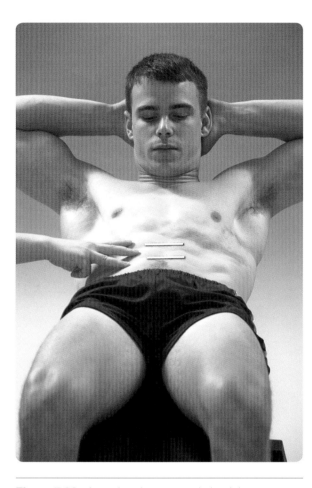

Figure 7.26 Locating the rectus abdominis.

Figure 7.27 Identifying the obliquus externus.

The rectus abdominis runs from the fifth, sixth, and seventh ribs and the xiphoid process of the sternum (see chapter 8) to the pubic symphysis (see later in this chapter) (figure 7.25). It flexes the lumbar and thoracic spine. The obliquus externus runs between the inferior edges of the last eight ribs and the outer edge of the middle half of the crest of the ilium. It flexes and contralaterally (to the opposite side) rotates the lumbar and thoracic spine. The obliquus internus runs between the outer edge of the middle two thirds of the crest of the ilium and the seventh, eighth, and ninth ribs. It flexes and ipsilaterally (to the same side) rotates the lumbar and thoracic spine. The transversus abdominis runs between the lower six ribs and the middle half of the internal edge of the crest of the ilium, the linea alba (formed by overlapping fascia of the abdominal muscles), and the pubic bone. Its primary function involving the spinal column is to assist with ipsilateral rotation of the thoracic and lumbar spine.

The posterior muscles of the spinal column are actually groups of muscles that cover two or more areas of the spinal column (figure 7.28). The spinalis group contains the **spinalis dorsi** (also called spinalis thoracis), the **spinalis cervicis,** and the **spinalis capitis.** The spinalis group of muscles originates from the second lumbar vertebra through the seventh cervical vertebra and inserts on the thoracic and cervical vertebrae through the occipital bone of the skull. Because their position is so closely aligned with the spinal column, their function is primarily extension. Another group of posterior spinal column muscles is the semispinalis, consisting of the **semispinalis dorsi** (or semispinalis thoracis), the **semispinalis cervicis,** and the **semispinalis capitis.** This group originates from the transverse processes of the seventh thoracic vertebra through the last four cervical vertebrae (C4–C7) and inserts on the spinous processes of the upper thoracic, all cervical vertebrae, and the occipital bone of the skull. The semispinalis group extends the cervical and thoracic spine and, particularly in the cervical section, is also involved in lateral flexion and rotation. A third group of posterior spinal column muscles is the iliocostalis group that includes the **iliocostalis lumborum,** the **iliocostalis dorsi** (or iliocostalis thoracis), and the **iliocostalis cervicis.** This group runs from the posterior aspect of the crest of the ilium and the 3rd through 12th ribs (lumborum and dorsi portions) and the transverse processes of the fourth, fifth, sixth, and seventh cervical vertebrae. All three sections of this group extend the spine, with the upper two sections also involved in lateral flexion and rotation. The fourth

group of posterior spinal column muscles is the longissimus group that includes the **longissimus capitis, longissimus cervicis,** and **longissimus dorsi** (or longissimus thoracis). This muscle group runs from the posterior aspect of the crest of the ilium and the transverse processes of the lumbar, thoracic, and lower cervical vertebrae to all thoracic vertebrae and cervical vertebrae up to the second. This group extends, laterally flexes, and rotates the lumbar, thoracic, and cervical sections of the spinal column.

Three additional muscles of the posterior aspect of the spinal column are the **sacrospinalis (erector spinae),** the **quadratus lumborum,** and the **mul-**

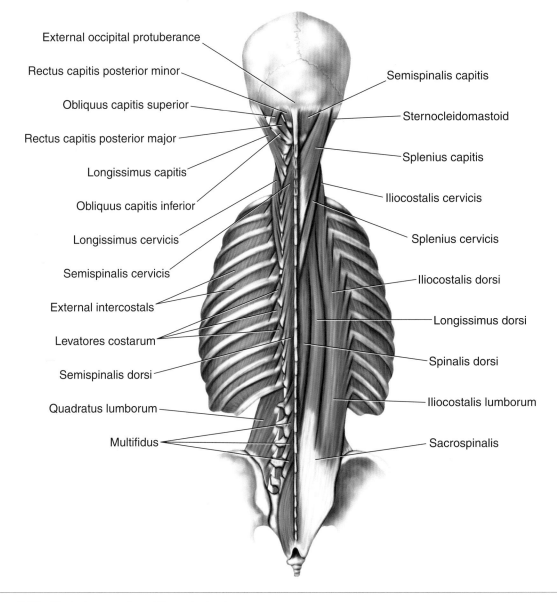

Figure 7.28   Posterior muscles of the spinal column.

tifidus. The sacrospinalis originates on the spines of the lumbar and sacral vertebrae and splits into three branches that are part of other muscle groups previously described: the iliocostalis lumborum, the longissimus dorsi, and the spinalis dorsi.

Hands on . . . Ask your partner to flex his or her trunk at the waist and hold this position. The erector spinae muscle group can be observed running parallel to the spinal column (figures 7.29 and 7.30).

These branches (and the sacrospinalis) extend and laterally flex the lumbar and thoracic spine. The quadratus lumborum runs from the posterior aspect of the crest of the ilium and transverse processes of the lower lumbar vertebrae to the transverse processes of the upper four lumbar vertebrae and the lower aspect of the 12th rib (figure 7.29). Although this muscle does contribute to the extension of the lumbar spine, its primary function is lateral flexion of the lumbar spine. The multifidus is a posterior muscle of the spinal column that literally covers all movable sections

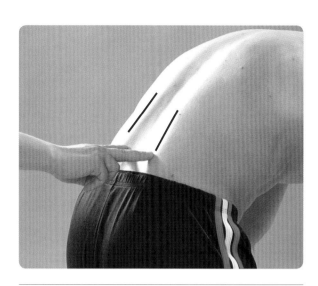

Figure 7.30   Finding the erector spinae group.

of the column. The multifidus originates from the sacrum; the posterior superior aspect of the iliac spine; and the lumbar, thoracic, and lower four cervical vertebrae and inserts on the spines of the lumbar, thoracic, and all but the first cervical vertebrae (see figure 7.28). This muscle extends and rotates the spinal column.

As did the cervical region of the spinal column, the thoracic and lumbar spine also have muscles with primary functions in the shoulder joint and shoulder girdle (figure 7.31). The trapezius (with

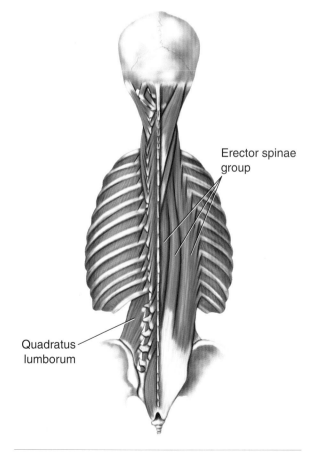

Erector spinae group

Quadratus lumborum

Figure 7.29   The erector spinae group, posterior view.

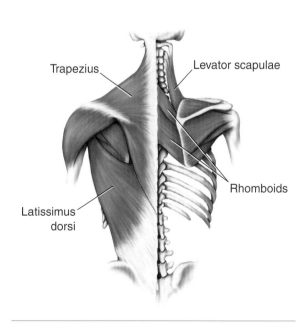

Trapezius

Levator scapulae

Latissimus dorsi

Rhomboids

Figure 7.31   Posterior view of the superficial muscles of the shoulder also acting on spinal column movement.

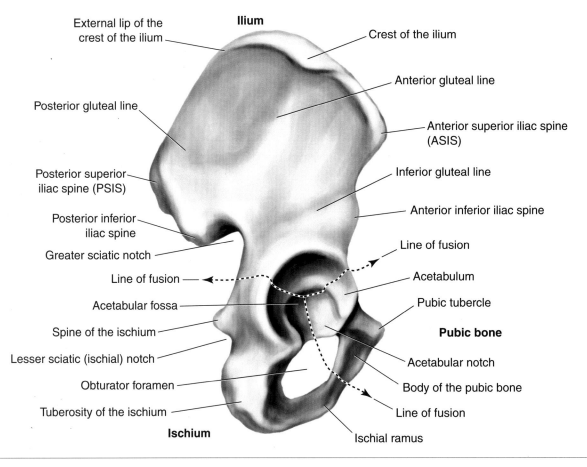

Figure 7.32   The innominate bone, lateral view (ilium, ischium, and pubic bone).

attachments to T1–T12), the rhomboids (with attachments to T1–T5), and the **latissimus dorsi** (with attachments to T1–T12 and L1–L5) also contribute to lateral flexion, rotation, and extension of the spinal column.

Bones of the Pelvis

Depending on your point of view, the pelvis could consist of 3, 7, or 11 bones (figure 7.32). The posterior aspect of the pelvis is the sacrum, the five sacral vertebrae fused together (figure 7.33). On both sides of the sacrum are two large bones often referred to as the **innominate bones.** Actually, each innominate bone is three separate bones that have fused together: the **ilium,** the **ischium,** and the **pubic bone.** Note the lines of fusion illustrated on the lateral view of the innominate bone in figure 7.32. The ilium bones articulate with the sacrum posteriorly and the pubic bones articulate with each other at the **pubic symphysis** to form the pelvis. The male and female pelvises have dif-

ferences to allow for childbearing by the female (figure 7.34). The female pelvis is proportionally wider and flatter and is tilted forward to a greater degree to allow for this function.

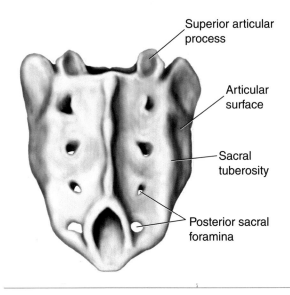

Figure 7.33   The sacrum, posterior view.

Male   **Female**

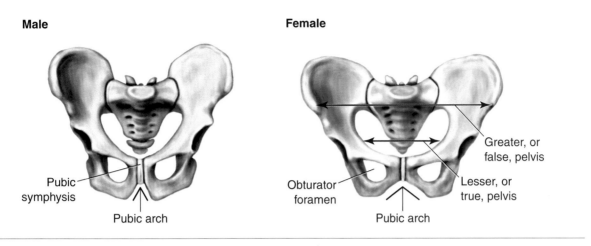

Figure 7.34 The differences between the male and female pelvis, anterior views.

The ilium, a wide, winglike structure, is the most superior part of the innominate bone and is the bone that articulates the pelvis with the spinal column through the sacrum.

At the most anterior and most posterior aspects of the ilium are bony prominences known as the **anterior superior iliac spine (ASIS)** and **posterior superior iliac spine (PSIS).** Running between the ASIS and PSIS is a large ridge of bone serving as a major source of muscular attachments known as the **crest of the ilium** (figure 7.32).

Hands on . . . Place your hands on your hips and feel your iliac crests (figure 7.35). While you are feeling your iliac crests, follow the crest to its anterior and posterior ends. These end points are your ASIS and PSIS (figure 7.36).

Beneath both the ASIS and PSIS are smaller bony prominences known as the **anterior inferior iliac spine** and the **posterior inferior iliac spine.** Just inferior to the posterior inferior iliac spine is a large notch known as the **greater sciatic notch.**

Figure 7.35 Finding the crest of the ilium.

Figure 7.36 Locating the posterior superior iliac spine.

Focus on . . . **THE CREST OF THE ILIUM**

The crest of the ilium is the source of many muscle attachments, both origins and insertions. When a person contuses this area, it is very painful to attempt to move the lower spinal column and the hips and even to attempt to breathe heavily because of the muscular attachments. This type of trauma to the soft tissue attaching to the crest of the ilium is often referred to as a "hip pointer."

Injuries such as a hip pointer can sometimes be the result of improperly fitted equipment or failure to use equipment. All coaches should have instruction in the proper use and fitting of equipment. Knowledge of specific anatomical structures and how equipment protects those structures can reduce the incidence and severity of some injuries.

The ischium is the most posterior of the three bones of the innominate bone and is distal to the ilium. Most posterior on the ischium is a bony prominence known as the **spine of the ischium.** Beneath the spine of the ischium is a notch in the bone known as the **lesser sciatic (ischial) notch.** At the very distal aspect of the ischium is a large bony prominence known as the **ischial tuberosity.** The ischial tuberosity serves as the source of attachment for the lower-extremity muscle group commonly known as the hamstrings.

Hands on . . . You sit on your ischial tuberosities. While standing, you can apply pressure to your gluteal areas and palpate your ischial tuberosities.

The pubic bone is the most anterior of the three bones making up the innominate bone. The pubic bones articulate with the ischium **(inferior pubic ramus),** the ilium **(superior pubic ramus),** and each other **(body of the pubic bone).** The articulation between the two pubic bones is known as the pubic symphysis (see figure 7.37). Just lateral to the pubic symphysis, each pubic bone has a bony prominence on its superior surface known as the **pubic tubercle.**

Hands on . . . Apply pressure at the most distal aspect of your abdomen and then move your hand distally to palpate your pubic symphysis.

On the lateral aspect of the innominate bone, the three bones (ilium, ischium, pubic bone) form a deep socket known as the **acetabulum.** This depression is the socket for the triaxial (ball-and-socket) hip joint. Beneath the acetabulum is a large foramen formed by the ischium and the pubic bone. This foramen is known as the **obturator foramen.**

Ligaments of the Pelvis

The **iliolumbar ligament** runs between the fifth lumbar vertebra and the crest of the ilium (figure 7.37). Two ligaments articulate the sacrum with the ilium: the **anterior sacroiliac** and the **posterior sacroiliac.** The anterior sacroiliac ligament runs between the anterior surface of the sacrum and the anterior surface of the ilium (figure 7.37). The posterior sacroiliac ligament (figure 7.37) has three sections: the **short sacroiliac,** the **long sacroiliac,** and the **interosseous.** The short sacroiliac ligament runs between the posterior ilium and the lower portions of the sacrum. The long sacroiliac ligament runs between the posterior superior spine of the ilium and the third and fourth vertebrae of the sacrum. The interosseous ligament is made of short fibers that connect the posterior aspects of the sacroiliac joint.

The ligaments of the pubic symphysis are the **anterior pubic,** the **inferior (arcuate) pubic,** the **posterior pubic,** and the **superior pubic.** The function of each of these ligaments is to articulate the anterior, posterior, superior, and inferior aspects of the two pubic bones to form the pubic symphysis. Additionally, there is an **interpubic fibrocartilage** disc between the pubic bones (figure 7.37).

Running between the anterior superior spine of the ilium to the pubic tubercle is a long ligament known as the **inguinal ligament,** which serves as a major source of muscular attach-

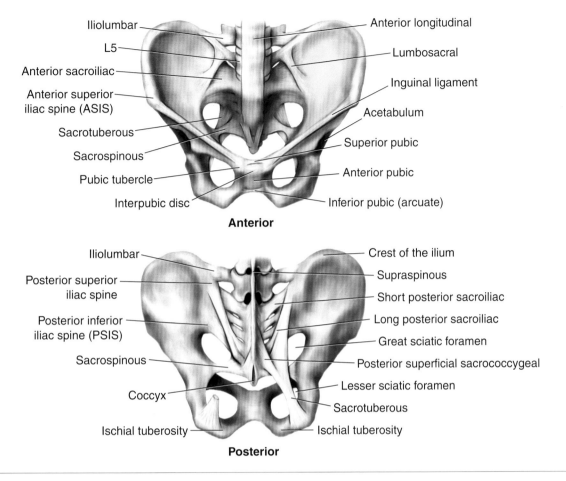

Figure 7.37 Landmarks and ligaments of the pelvis.

ments (figure 7.37). Two ligaments that stabilize the pelvis are the **sacrospinous** and the **sacrotuberous** ligaments (figure 7.37). The sacrospinous ligament runs from the sacrum and coccyx to the spine of the ischium. The sacrotuberous ligament runs between the posterior inferior spine of the ilium, the sacrum and coccyx, and the ischial tuberosity.

Movements and Muscles of the Pelvis

Movement is not normal in the joints of bones in the pelvis. The sacroiliac joint and the joints between the ilium, the ischium, and the pubic bone are essentially fused. The pubic symphysis has minimal movement of a gliding nature.

Focus on . . . INGUINAL HERNIA

Excessive stress on the muscles that attach to the inguinal ligament can result in an *inguinal hernia.* The term *hernia* indicates an abnormal protrusion of an organ. Straining the musculature that attaches to the inguinal ligament can cause a portion of the abdominal contents to protrude through the strained muscle tissue, resulting in an inguinal hernia.

Improper lifting techniques can cause hernias and are also a leading contributor of low back muscular strain. Proper lifting techniques can be taught in weight-training programs and in academic courses in kinesiology and biomechanics.

A combination of movements of the spinal column and the hip joint results in pelvic motion: backward tilt, forward tilt, lateral tilt, and rotation. Backward tilt of the pelvis involves the pubic symphysis moving upward and the sacrum moving downward. This is accomplished by flexion of the lumbar spinal column and extension of the hip joints (figure 7.38b). Forward tilt of the pelvis involves the pubic symphysis moving downward and the sacrum moving upward. Forward pelvic tilt is accomplished by extension of the lumbar spinal column and flexion of the hip joints (figure 7.38c). Lateral tilt of the pelvis results when one ilium is elevated above the other ilium. This is accomplished by lateral flexion of the lumbar spinal column, abduction of one hip joint, and adduction of the other hip joint (figure 7.39). Rotation of the pelvis involves rotation of the lumbar spine with internal rotation of one hip joint and external rotation of the other hip joint (figure 7.40).

The muscles of the spinal column were presented in this chapter, and the muscles of the hip joint are presented in chapter 10. The spinal column muscles produce the flexion, extension, lateral flexion, and rotation necessary for the pelvic movements, along with the muscles that produce flexion, extension, abduction, adduction, and internal and external rotation of the hip joint. From this information you should be able to understand which muscles produce which movements of the pelvis through spinal column and hip joint movements.

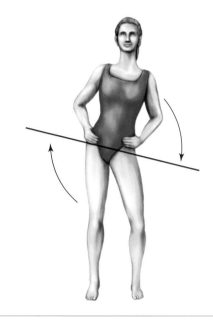

Figure 7.39   Lateral tilt of the pelvic girdle.

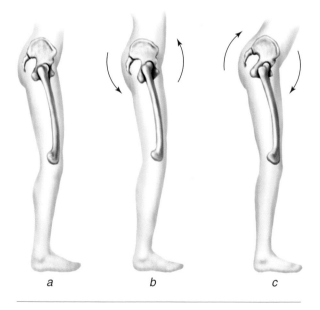

a   b   c

Figure 7.38   Movements of the pelvic girdle: *(a)* anatomical position, *(b)* backward tilt, and *(c)* forward tilt.

Figure 7.40   Rotation of the pelvic girdle.

REVIEW OF TERMINOLOGY

The following terms were discussed in this chapter. Define or describe each term, and where appropriate, identify the location of the named structure either on your body or in an appropriate illustration.

acetabulum
annulus fibrosus
anterior inferior iliac spine
anterior longitudinal ligament
anterior pubic ligament
anterior sacroiliac ligament
anterior superior iliac spine (ASIS)
atlantoaxial ligaments
atlanto-occipital ligaments
atlas
axis
bifid
body of the pubic bone
cervical vertebrae
coccyx
crest of the ilium
dens
erector spinae
greater sciatic notch
iliocostalis cervicis
iliocostalis dorsi
iliocostalis lumborum
iliolumbar ligament
ilium
inferior (arcuate) pubic ligament
inferior pubic ramus
inguinal ligament
innominate bones
interbody ligaments
interosseous ligament
interpubic fibrocartilage
interspinous ligament
intervertebral disc

ischial tuberosity
ischium
kyphosis
latissimus dorsi
lesser sciatic (ischial) notch
levator scapulae
ligamentum flavum
ligamentum nuchae
longissimus capitis
longissimus cervicis
longissimus dorsi
long sacroiliac ligament
longus capitis
longus colli
lordosis
lumbar vertebrae
multifidus
nucleus pulposus
obliquus capitis superior and inferior
obliquus internus and externus
obturator foramen
occipitoaxial ligaments
odontoid process
posterior inferior iliac spine
posterior longitudinal ligament
posterior pubic ligament
posterior sacroiliac ligament
posterior superior iliac spine (PSIS)
prevertebral muscles
pubic bone
pubic symphysis
pubic tubercle
quadratus lumborum

rectus abdominis
rectus capitis anterior
rectus capitis lateralis
rectus capitis posterior major and
   minor
rhomboids
sacral vertebrae
sacrospinalis (erector spinae)
sacrospinous ligament
sacrotuberous ligament
sacrum
scalenus anterior
scalenus medius
scalenus posterior
scoliosis
semispinalis capitis
semispinalis cervicis
semispinalis dorsi
short sacroiliac ligament
spinalis capitis
spinalis cervicis
spinalis dorsi
spine of the ischium
splenius capitis
splenius cervicis
sternocleidomastoid
superior pubic ligament
superior pubic ramus
supraspinous ligament
thoracic vertebrae
transverse ligament
transversus abdominis
trapezius

SUGGESTED LEARNING ACTIVITIES

1. Imagine you are wearing a jacket with slash-type pockets on both sides of the front of the jacket just above your waist. Place your hands inside those pockets.

   a. What direction (horizontally, vertically, diagonally) are your fingers pointing?
   b. What abdominal muscle fibers are parallel to the direction your fingers are pointing?

2. In a tackling technique known in football as "clotheslining," a player's head is stopped while the rest of his body continues in motion. This could cause paralysis or even death. Describe anatomically what happens in the cervical spine and how this might have catastrophic results.

3. Excessive weight gain, possibly a pregnancy, or other causes may shift one's center of gravity forward. This could cause the lumbar spine to develop an excessive curvature to counteract the shift.

   a. In which direction would the lumbar spine curve?
   b. What would this excessive curvature be called?

4. Imagine yourself as a baseball pitcher, a football quarterback, a mail carrier with a heavy pouch, a golfer, a tennis player, or any person who participates in any activity that requires use of the musculature on one side of the body more than the other.

(continued)

SUGGESTED LEARNING ACTIVITIES   *(continued)*

What type of abnormal curvature of the spine might such activity develop?

5. Name the muscles responsible for turning your head to the right from the anatomical position.

6. Lying supine (on your back), draw your thighs up to your chest.

    a. What movement occurred in your lumbar spine?

    b. What movement occurred in your hip joints?

    c. What movement occurred in your pelvis?

MULTIPLE-CHOICE QUESTIONS

1. How many vertebrae are in the cervical spine?
    a. 4
    b. 5
    c. 7
    d. 12

2. How many vertebrae are in the thoracic spine?
    a. 4
    b. 5
    c. 7
    d. 12

3. How many vertebrae are in the lumbar spine?
    a. 4
    b. 5
    c. 7
    d. 12

4. How many vertebrae are in the coccygeal spine?
    a. 4
    b. 5
    c. 7
    d. 12

5. When a person shakes his or her head no, the primary movement takes place between
    a. the skull and C1
    b. C1 and C2
    c. C1 and C7
    d. C2 and C7

6. Which of the following bones of the pelvic girdle is not part of the structure known as the acetabulum?
    a. ilium
    b. ischium
    c. sacrum
    d. pubic bone

7. A forward tilt of the pelvic girdle requires the pubic symphysis to
    a. move laterally
    b. move downward
    c. rotate
    d. move upward

8. A forward tilt of the pelvic girdle requires the sacrum to
    a. move laterally
    b. move downward
    c. rotate
    d. move upward

9. A backward tilt of the pelvic girdle requires the pubic symphysis to
    a. move laterally
    b. move downward
    c. rotate
    d. move upward

10. A backward tilt of the pelvic girdle requires the sacrum to
    a. move laterally
    b. move downward
    c. rotate
    d. move upward

11. Which of the following is not considered a fundamental movement of the vertebral column?
    a. flexion
    b. abduction
    c. extension
    d. rotation

12. Which of the following bones does not have an attachment of the sternocleidomastoid muscle?
    a. sternum
    b. clavicle
    c. mastoid
    d. humerus

13. Considering that there are right- and left-side scalene muscles, under normal conditions, how many scalene muscles total are there in the cervical spine?
    a. 2
    b. 3
    c. 4
    d. 6

14. The erector spinae muscle group supports various segments of the spinal column and is located in what direction relative to the spinal column?
    a. anterior
    b. posterior
    c. superior
    d. inferior

15. Running from the pubic bone to the ribs, which of the following muscles is the most likely to be exclusively a flexor of the lumbar spine?
    a. rectus abdominis
    b. transversus abdominis
    c. obliquus externus
    d. obliquus internus

16. Normally, the cervical spine has what type of curvature?
    a. anterior
    b. posterior
    c. lateral
    d. medial

FILL-IN-THE-BLANK QUESTIONS

1. Under normal conditions, the spinal column has anterior curves in the cervical spine and the _____ _____ spine.

2. Under normal conditions, the spinal column has posterior curves in the coccygeal spine and the _____ spine.

3. The ligament of the spinal column that is considered to be the anterior wall of the spinal canal is the _____ ligament.

4. The cervical vertebra attached to the skull is known as the _____.

5. The anatomical structure that serves as a shock absorber between bodies of the spinal vertebrae is known as the _____.

6. The anterior connection of the two innominate bones is known as the _____.

7. The hamstring muscles originate on the large tuberosity of the _____.

8. An excessive anterior curvature of the lumbar spine is known as _____.

9. An excessive posterior curvature of the thoracic spine is known as _____.

10. An excessive lateral curvature of the spinal column is known as _____.

The Thorax

The thorax (figure 8.1) is a structure formed by bones that create a large compartment known as the thoracic cavity (chest cavity), which houses the lungs and heart. Linings of the lungs and the inner walls of the cavity create a vacuum, and movement of the thorax creates changes in pressure within the thoracic cavity, which causes air to enter into or be expelled from the lungs. Normal inspiration and expiration of air by the lungs are commonly known as quiet respiration, whereas movement of air in and out of the lungs during physical exertion is known as forced respiration. Movement at the joints of the thorax allows for the expansion and contraction of the thoracic cavity, and muscles attached to the bones of the thorax create that movement.

Bones of the Thorax

The anterior bone of the thorax is known as the **sternum** (figure 8.2). The sternum protects the structures beneath it; serves as a source of muscular attachment for muscles of the thorax, the neck, and the abdomen; and provides attachment for costal cartilage. The sternum consists of three parts: the **manubrium,** the **body,** and the **xiphoid process.** The manubrium has a **suprasternal** (jugular) **notch** at its superior edge (figure 8.3). The line where the inferior edge of the manubrium attaches to the body of the sternum is known as the angle of the sternum. The manubrium also has two **clavicular notches** and two **first costal notches** where the sternum articulates with the clavicles (sternoclavicular joints) and the first ribs of the thorax. The body of the sternum has costal notches along the lateral sides where the ribs articulate. The most inferior portion of the sternum, the xiphoid process, is cartilaginous in early life and becomes bone in the adult.

The thorax has 12 pairs of **ribs:** The first (superior) seven pairs are known as true ribs and the last (inferior) five pairs are known as false ribs. The

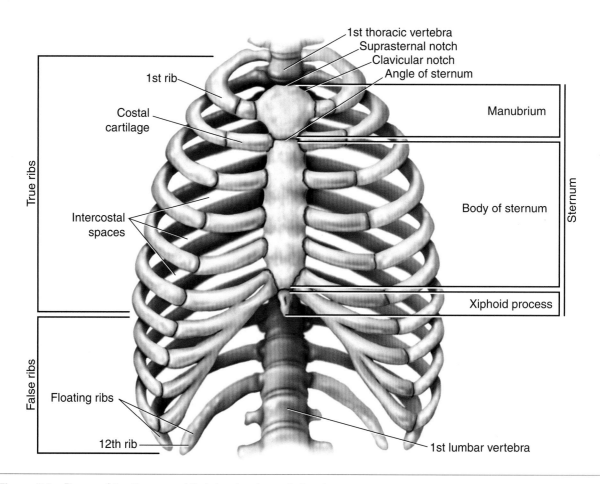

Figure 8.1   Bones of the thorax and their landmarks, anterior view.

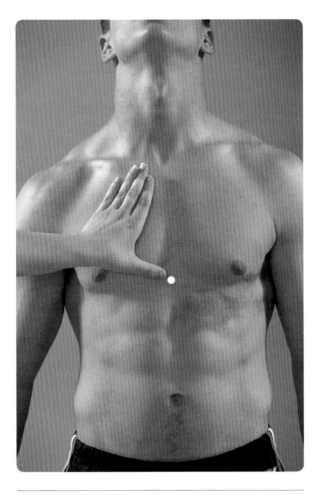

Figure 8.2   Locating the sternum.

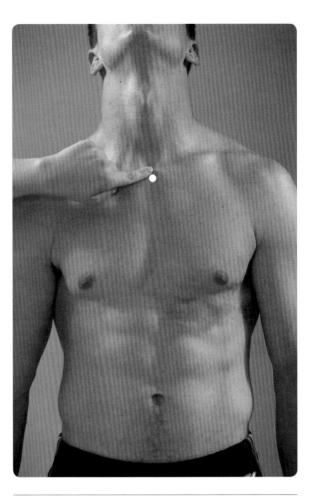

Figure 8.3   Locating the suprasternal notch.

first seven ribs attach to the sternum (the first to the manubrium and the second through seventh to the sternal body) through costal cartilage, which extends from 1 to 3 in. (2.5–7.6 cm), depending on the length of the rib, between the sternal end of the

rib and the sternum. The first three false ribs (ribs 8–10) are attached indirectly to the sternum by costal cartilage that attaches to the costal cartilage of the rib above it. The last two pairs of false ribs are also referred to as floating ribs because they

---

Focus on . . . CPR AND THE XIPHOID PROCESS

When learning cardiopulmonary resuscitation (CPR), the rescuer is taught to use the xiphoid process as a landmark to establish hand placement on the body of the sternum for compressions. Pressure applied to the xiphoid process not only produces ineffective compressions but also creates the possibility of fracture of the xiphoid process. First-aid courses; CPR courses; emergency medical technician/paramedic courses; and programs in medicine, nursing, physical therapy, athletic training, and other allied health sciences require the acquisition of CPR skills. Knowledge of the anatomy of the thorax helps you to perform these CPR skills properly and not expose the victim to further trauma.

have no anterior attachment. The **head** and the **neck** of a rib (figure 8.4), with an articular facet and tubercle, appear at the end of the rib that articulates with a thoracic vertebra. Lateral to the head of the rib is the angle. This creates the broad back of the human body and gives humans the ability to lie down in a supine position without rolling to one side or the other.

In addition to the sternum and the 12 ribs, the 12 thoracic vertebrae of the spinal column are also considered bones of the thorax. The anatomy of the thoracic vertebrae is covered in chapter 7, The Spinal Column and Pelvis

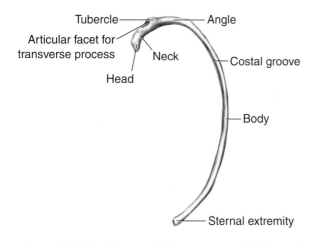

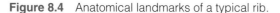

Figure 8.4   Anatomical landmarks of a typical rib.

Joints and Ligaments of the Thorax

The anterior articulations of the thorax (figure 8.5) include the **chondrosternal joints,** the **costochondral joints,** and the **interchondral joints.** The chondrosternal joints articulate the costal cartilages of the upper seven (true) ribs with the costal notches of the manubrium (first rib) and the body of the sternum (ribs 2–7). The ligaments include the **costochondral (capsular),** the **costosternal,** and the **interchondral (interarticular).** A gliding, rotating motion occurs at these joints, which allows external rotation and elevation of the rib. The costochondral joints are the articulations between the sternal end of the rib and the cartilage attaching the rib to the sternum (ribs 1–7) or to the cartilage of the rib above (ribs 8–10). There is no actual ligament associated with the costochondral articulation because the periosteum of the rib (a covering of connective tissue) in this case is actually continuous with the perichondrium of the cartilage (a thin, fibrous tissue that covers cartilage and provides it with nutrition). In athletics, the costochondral joint often comes under stress when the thorax is compressed beyond the normal range of motion of this joint. Slight rotation of the costochondral joint allows upward and outward movement of the rib.

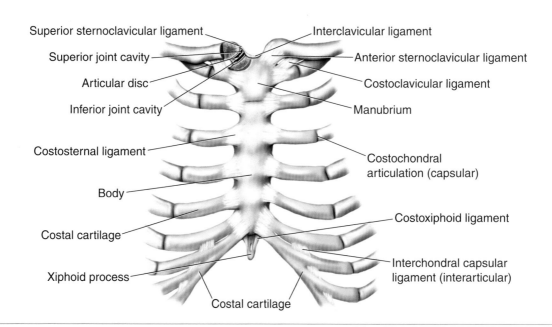

Figure 8.5   The costochondral joint and anterior ligaments of the thorax.

Focus on . . . **COSTOCHONDRAL SPRAIN**

The costochondral joint, like other joints, has a normal range of motion. When that range of motion is exceeded, a costochondral sprain is possible. The thorax of a football player being tackled may be compressed between the tackler and the playing surface, forcing the costochondral joints beyond their normal limits. Likewise, the thorax of a wrestler taken down from a standing position may be compressed between the opponent and the wrestling mat, causing injury to the costochondral joints. Some old-time wrestling coaches refer to this injury as a "rib out." If the costochondral articulation is sprained, the rib and the cartilage may separate, allowing the bone to move more than normal and resulting in the appearance of a rib attempting to come through the skin at the site of the sprain. The mechanism that causes a costochondral sprain can result in other forms of trauma also (e.g., rib fractures, damage to internal organs of the thoracic cavity).

Recognizing a costochondral sprain requires study of human anatomy, athletic training, sports medicine, and coaching. Care of such a sprain is essential for proper healing. Although it would be ideal to completely immobilize the joint for several weeks (i.e., prevent breathing), this is obviously not possible. Other means of restricting movement need to be used, such as rib belts or taping or wrapping, that will reduce movement of the costochondral articulation during respiration.

All 12 ribs articulate with the 12 thoracic vertebrae through two articulations for each rib. The **costovertebral articulations** are known as the **corpocapite** and **costotransverse capsular articulations.** The corpocapite articulation is between the head of the rib and the body of the adjacent thoracic vertebra, and the ligaments involved are the **costotransverse capsular** and the **costovertebral radiate** (figure 8.6).

A gliding, rotational movement allows elevation and depression of the ribs. The costotransverse articulations are the joints between the tubercles of the first 10 ribs with the adjacent vertebra's transverse process. The 11th and 12th ribs do not have an articulation with the adjacent transverse processes. The only ligament of these costotransverse articulations is the capsular ligament.

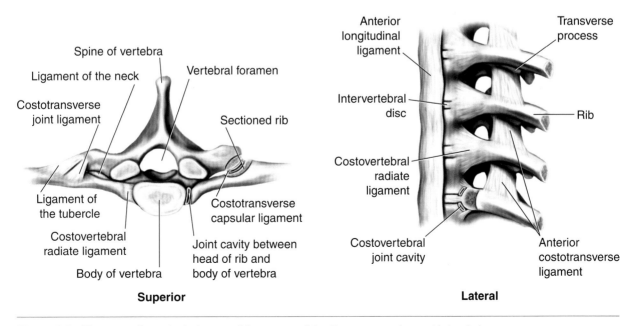

Figure 8.6   The posterior articulations and ligaments of the thorax, superior and lateral views.

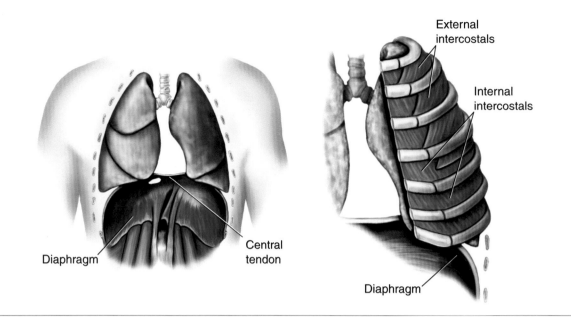

Figure 8.7   Main respiratory muscles.

Movements and Muscles of the Thorax

Respiration depends on movements of the thorax. As mentioned earlier, normal respiration is known as quiet respiration. The muscles of quiet respiration are considered specific to respiration (figure 8.7). Muscles discussed in other chapters become involved in forced respiration if they have any attachment to the bones of the thorax.

The muscles of the thorax include the **diaphragm.** This large, domelike muscle runs from the xiphoid process, cartilage of the last seven ribs, and lateral aspect of the first four lumbar vertebrae to a structure known as the **central tendon** (figure 8.8). When the diaphragm contracts, its central tendon pulls downward, increasing space in the thoracic cavity and thus changing the pressure within the cavity. As air is drawn into the lungs and they expand, both the **internal** and **external intercostal muscles** (figure 8.9) draw the ribs together, allowing the lungs to expand outward and downward.

Hands on . . . Raise either arm above your head; place your fingers in the space between the anterior lateral aspect of the thorax to place pressure on the intercostal muscles (figure 8.10). The external intercostals (11 pairs), on the anterior aspect of the ribs, run between the lower border of a rib to the upper border of the rib below. Note

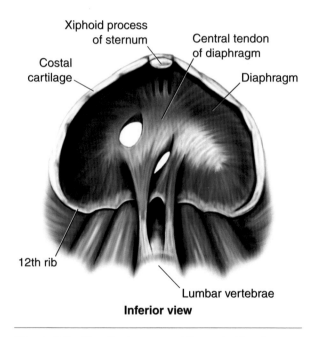

Inferior view

Figure 8.8   The diaphragm and its central tendon.

the angle made by the direction of the external intercostal fibers.

Hands on . . . Place your open hands in the pockets of a jacket. Note the direction your fingers are pointing. Is there any similarity between the angle your fingers point and the angle the external intercostal fibers run?

The internal intercostals (11 pairs), on the posterior aspect of each rib, run between the upper

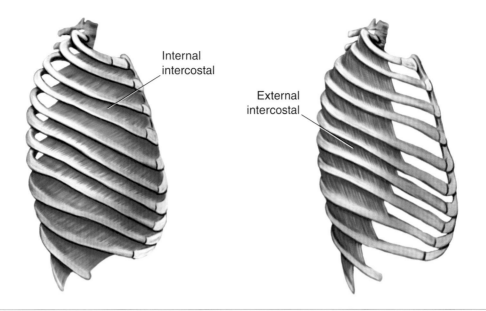

Figure 8.9 External and internal intercostal muscles, lateral view.

border of a rib to the upper border of the rib below. The movement created by contraction of both the external and internal intercostal muscles is a drawing together of the ribs. If the scalene muscles of the cervical spine contract to fix (stabilize) the first rib, the external intercostal muscles elevate the ribs to allow more volume in the thoracic cavity for the lungs to fill. If the last rib is fixed by the quadratus lumborum muscle, the lower ribs are drawn further together to decrease the volume of the thoracic cavity.

Other thorax muscles include the **levatores costarum** (12 pairs), which run from the transverse processes of the vertebrae C7 through T11 to the angle of the rib below (figure 8.11). Contraction assists with elevation of the rib. The **subcostal muscles** (10 pairs) run between the inner surface of a rib near its angle to the inner surface of the second or third rib below (figure 8.12). Contraction assists with elevation of the rib. The **serratus posterior** muscle has inferior and superior portions (figure 8.13). The inferior portion runs from the spinous processes of the last two thoracic and first three lumbar vertebrae to the inferior border of the last four ribs just lateral to their angles and depresses these ribs. The superior portion of the serratus posterior runs from the spinous processes of the seventh cervical and first three thoracic vertebrae to just lateral to the angle of the second, third, fourth,

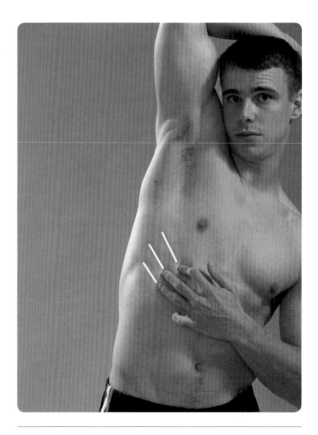

Figure 8.10 Locating the intercostals.

and fifth ribs and elevates these ribs. The last muscle considered a muscle of the thorax is the **transversus thoracis** (figure 8.14). It runs from the lower third of the posterior aspect of the sternum,

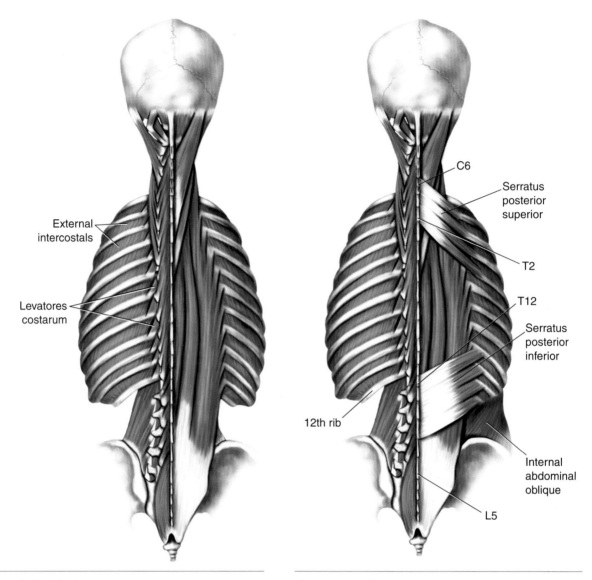

Figure 8.11   The levatores costarum.

Figure 8.13   The serratus posterior, superior and inferior portions.

the xiphoid process, and the costal cartilage of the fourth through seventh ribs to the inner aspect of the costal cartilage of the second to sixth ribs and elevates the ribs.

Numerous other muscles have one of their attachments to one of the bones of the thorax and usually are not involved in quiet respiration but can be involved in forced respiration, when lung capacity needs to increase to meet demands and therefore needs additional space within the thoracic cavity to expand. Table 8.1 describes the muscles involved in quiet respiration and those involved in forced respiration.

The information in table 8.1 implies that **quiet**

inspiration requires the diaphragm to contract, and as the lungs fill with air, the external and internal intercostal muscles contract to elevate the ribs to create more space for the expanded lungs. When the diaphragm and intercostal muscles relax and return to their starting positions, **quiet expiration** occurs. In **forced inspiration,** to make space for greater filling of the lungs, additional muscles that have an attachment to the thorax and are capable of expanding the thorax are called into action. When as much air as possible needs to be expelled from the lungs, any muscle capable of squeezing the thoracic cavity as small as possible is called into action during **forced expiration.**

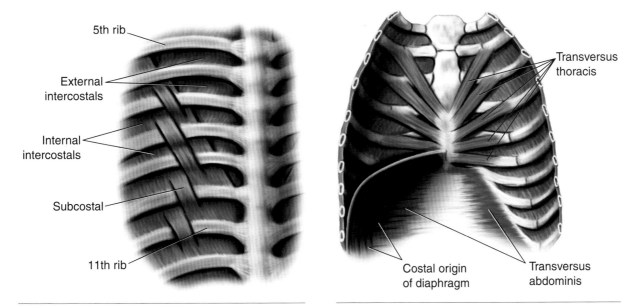

Figure 8.12 The subcostal muscles.

Figure 8.14 The transversus thoracis.

Table 8.1 Muscles of Respiration

Muscle	Quiet respiration		Forced	
	Inspiration	**Expiration**	**Inspiration**	**Expiration**
Diaphragm	+	Passive	+	
External intercostals	+	Passive	+	+
Internal intercostals	+	Passive	+	+
Latissimus dorsi			+	
Levatores costarum			+	
Pectoralis major and minor			+	
Rhomboids major and minor			+	
Scaleni anterior, middle, and posterior			+	
Serratus anterior			+	
Serratus posterior (superior aspect)			+	
Sternocleidomastoid			+	
Subcostals			+	
Trapezius			+	
Obliqui externus and internus			+	
Rectus abdominis				+
Sacrospinalis group				+
Serratus posterior (inferior aspect)				+
Transversus abdominis				+
Transversus thoracis				+

REVIEW OF TERMINOLOGY

The following terms were discussed in this chapter. Define or describe each term, and where appropriate, identify the location of the named structure either on your body or in an appropriate illustration.

body of the sternum
central tendon
chondrosternal joints
clavicular notches
corpocapitate articulation
costochondral (capsular) ligament
costochondral joints
costosternal ligament
costotransverse capsular articulation
costotransverse capsular ligament
costovertebral articulations

costovertebral radiate ligament
diaphragm
first costal notches
forced expiration
forced inspiration
head of a rib
interchondral (interarticular) ligament
interchondral joints
internal and external intercostal
    muscles
levatores costarum

manubrium
neck of a rib
quiet expiration
quiet inspiration
ribs
serratus posterior
sternum
subcostal muscles
suprasternal notch
transversus thoracis
xiphoid process

SUGGESTED LEARNING ACTIVITIES

1. On your body or your partner's body, trace the clavicle with your finger from its most lateral end to its medial end.
   a. At the medial end of both clavicles, what do you feel?
   b. What is the name of the area you are palpating?

2. Raise your or your partner's hand and arm laterally overhead. Find the clavicle, and palpate all 12 ribs.
   a. Which, if any, ribs were difficult or impossible to palpate?
   b. Why?

3. With a cloth tape measure, determine the circumference of your partner's thorax.
   a. After determining the resting (quiet) circumference, have your partner inhale to the greatest degree possible, and then remeasure the circumference. Then have your partner forcefully exhale all the air possible from the lungs, and again remeasure the thorax circumference. (Note: Be sure to measure each time at the same level of the thorax.)
   b. What muscles caused the thorax to expand?
   c. What muscles caused the thorax to contract?

4. Assume the position for applying chest compressions as in cardiopulmonary resuscitation (CPR). *Do not actually apply this technique!*
   a. Place the heel of your hand on your partner's sternum. Identify the manubrium, the body, and the xiphoid process.
   b. Identify the thorax joints that allow you to compress the thorax without fracturing ribs.

MULTIPLE-CHOICE QUESTIONS

1. How many pairs of ribs are typically referred to as false ribs?
   a. 2
   b. 5
   c. 7
   d. 12

2. The ligament connecting the bony portion of a rib to the costal cartilage is known as the
   a. costosternal
   b. costochondral
   c. costoclavicular
   d. costothoracic

3. Quiet expiration requires action by which of the following muscles?
   a. none
   b. diaphragm
   c. intercostals
   d. scaleni

FILL-IN-THE-BLANK QUESTIONS

1. The thorax consists of thoracic vertebrae, ribs, costal cartilage, and the _____.

2. The superior portion of the sternum is known as the _____.

3. The chief muscle of respiration is known as the _____.

Nerves and Blood Vessels of the Spinal Column and Thorax

CHAPTER OUTLINE

As in the upper and lower extremity, the nerves innervating the thorax and trunk originate from plexuses (networks) of nerves from the spinal cord. In particular, one plexus containing a nerve that innervates the lungs is vital to respiration and the maintenance of life. Additionally, the major nerves that serve the entire lower extremity originate at the lower end of the spinal column (lumbosacral plexus). The blood vessels (arteries and veins) that provide circulation to the thorax and the trunk are housed within the thoracic and abdominal cavities. This chapter concentrates on the nerves and blood vessels that supply the muscles of the thorax, trunk, and respiration. These structures are identified in the summary table found at the end of this section for the spinal column, pelvis, thorax, and trunk.

Nerves of the Thorax and Trunk

Spinal nerves innervating the musculature are formed into plexuses, or networks of nerves. The nerves that innervate the head and cervical spine arise from the **cervical plexus** (figure 9.1). The sensory nerves of the plexus are superficial, whereas the motor nerves are the deep branches. The deep branches innervate the muscles of the scalp and face and also the cervical spine muscles, such as the sternocleidomastoid, the trapezius, and the levator scapulae. Branches from a cranial nerve, the **spinal accessory,** also communicate with C2, C3, and C4 and innervate the sternocleidomastoid and trapezius muscles. Additional deep branches of the cervical plexus innervate the rectus capitis anterior and lateralis muscles, the longus capitis and colli muscles, the prevertebral muscles, the levator scapulae, the scalenus medius muscle, the sternocleidomastoid, and the trapezius. The most important nerve of the cervical plexus for respiration is the **phrenic nerve** (C3, C4, C5), which innervates the **diaphragm.** Other nerves of the brachial plexus that innervate the muscles of the cervical spine and lateral aspect of the thorax are described in chapter 6.

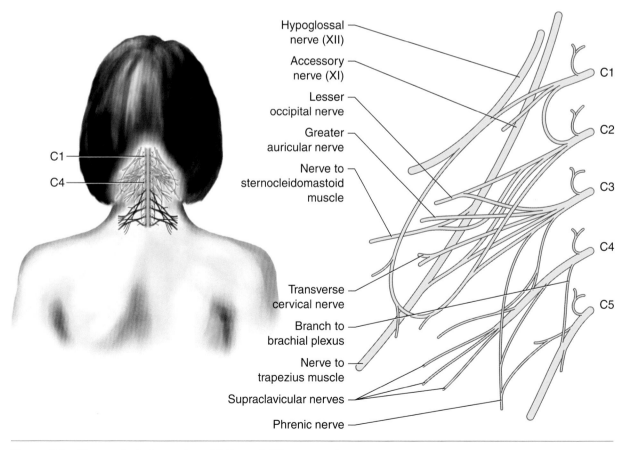

Figure 9.1   The cervical plexus, from C1 through C5 nerve roots.

Focus on . . . CERVICAL SPINE INJURY

When an individual suffers a cervical spine injury, it is extremely important to assess the person's vital signs, including respiration. Trauma to the phrenic nerve can disrupt breathing, necessitating the first-aid technique known as *rescue breathing*. Disruption of breathing combined with cessation of the pulse calls for the initiation of cardiopulmonary resuscitation (CPR).

There are 12 **thoracic nerves,** often described as 11 **intercostal nerves** and one **subcostal nerve,** that arise from the thoracic section (T1–T12) of the spinal column (figure 9.2). The first thoracic intercostal nerve innervates the levatores costarum muscles, intercostal muscles, and superior portion of the serratus posterior muscles. The second thoracic intercostal nerve innervates the intercostal muscles, the levator costarum muscles, and the subcostal muscles. The third thoracic intercostal nerve innervates the levator costarum, intercostal, and subcostal muscles. The fourth thoracic intercostal nerve innervates the levator costarum, intercostal, subcostal, and transversus thoracis muscles. The fifth and sixth thoracic intercostal nerves innervate the intercostal, subcostal, and transversus thoracis muscles. The seventh and eighth thoracic intercostal nerves innervate the intercostal muscles. The 9th, 10th, and 11th thoracic intercostal nerves innervate the intercostals, the inferior portion of the serratus posterior, and the abdominal muscles. The 12th thoracic subcostal nerve innervates the abdominal muscles.

The **lumbosacral plexus** is formed by the anterior rami of the nerves of the **lumbar plexus,** the **sacral plexus,** and the **pudendal** (coccygeal) **plexus** (figure 9.2). Most of these nerves innervate the structures of the lower extremity–including the hips, buttocks, groin, and organs of the pelvic region–and are therefore discussed in chapter 13 on the nerves and blood vessels of the lower extremity.

Arteries of the Thorax and Trunk

The **ascending aorta** arises from the heart and has right and left **coronary arteries,** which have branches to the heart itself. The area between the ascending and **descending aorta** is known as the **aortic arch** (figure 9.3), which first gives rise to the **brachiocephalic artery** that supplies blood

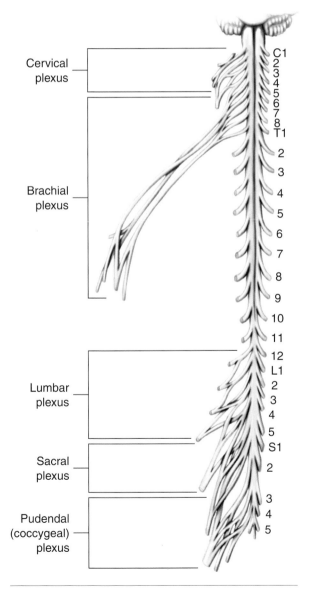

Figure 9.2   The spinal nerves and major plexuses.

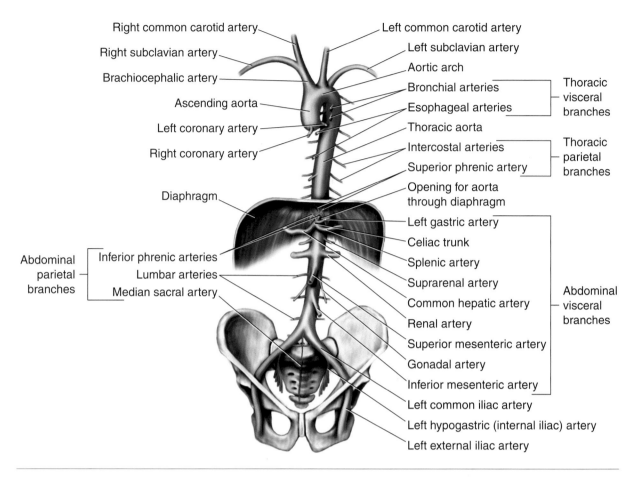

Figure 9.3   Major arteries of the trunk, thorax, and abdomen.

to the right arm, the thorax, and the right side of the head and neck. The first branches from the brachiocephalic artery are the right **common carotid** and right **subclavian arteries.** The subclavian arteries were discussed in chapter 6 on nerves and blood vessels of the upper extremity. The right common carotid artery, just below the jaw, divides into the **external** and **internal carotid arteries.**

Hands on . . . To obtain a carotid artery pulse, place your index and middle fingers on your Adam's apple and move them between this structure and the anterior muscles of your neck (figure 9.4).

The external carotid artery and its branches supply the muscles of the head and upper portion of the neck. The internal carotid artery and its branches supply the brain and other structures of the head. Additionally, the **vertebral arteries,** arising from the subclavian arteries, supply the brain, upper portions of the spinal cord, cervical vertebrae, and deep muscles of the neck. An artery between the two vertebral arteries known as the **basilar artery** and its branches (the **posterior cerebral arteries**) supply structures within the skull.

The descending aorta, initially designated as the **thoracic aorta,** has both **thoracic** and **abdominal parietal** and **thoracic** and **abdominal visceral branches** (figure 9.3). The thoracic parietal branches (intercostal, subcostal, superior phrenic) supply muscles such as the intercostals, the vertebral column muscles, and the diaphragm. The thoracic visceral branches (bronchial, esophageal) supply the structures within the thoracic cavity. Below the diaphragm, the descending aorta becomes known as the **abdominal aorta** (figure 9.3). The abdominal parietal branches (inferior phrenic,

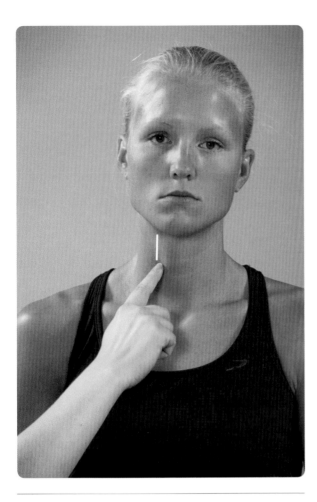

Figure 9.4   Finding the carotid artery.

lumbar, median sacral) supply the diaphragm and the musculature of the abdominal, lumbar, and sacroiliac areas. The abdominal visceral branches (gastric, celiac, splenic, suprarenal, renal, common hepatic, superior and inferior mesenteric, gonadal) supply the organs of the abdominal cavity, including the liver, spleen, stomach, suprarenals, kidneys, pancreas, intestines, and reproductive structures.

At approximately the L4 level of the vertebral column, the abdominal aorta divides into the right and left **common iliac arteries** (figure 9.3). The common iliac artery divides into the **external** and **internal iliac (hypogastric) arteries.** The branches supply structures in the area of the sacrum and coccyx as well as the iliacus muscles.

The arteries in the lower abdominal cavity and pelvic region are divisions of the common iliac artery mentioned earlier: the external iliac and the hypogastric (internal iliac) arteries. The external iliac artery supplies the abdominal muscles, the iliacus, the psoas muscles, the sartorius, and the tensor fascia lata of the lower extremity. At the level of the inguinal ligament, the external iliac artery becomes known as the **femoral artery,** which is discussed further in chapter 13 on the nerves and blood vessels of the lower extremity. The internal iliac artery has muscular branches that supply muscles of the pelvic girdle and the hip joint.

Veins of the Thorax and Trunk

With a few exceptions, the veins of the thorax and trunk (figure 9.5) parallel the arteries they drain and also have the same or similar names. The **superior vena cava** is the major vein that drains the head, neck, shoulders, upper extremity, and parts of the thorax and abdomen into the heart. The **brachiocephalic vein** receives blood from the subclavian veins (discussed in chapter 6 on nerves and blood vessels of the upper extremity) and drains into the superior vena cava. The veins draining the head and neck include the **internal jugular,** the **vertebral,** and the **external jugular veins** and their branches. The internal jugular drains the face and structures of the throat into the subclavian vein. The vertebral vein drains the structures of the skull, posterior neck muscles, and the first intercostal area into the brachiocephalic vein. The external jugular and its more superficial branches drain the superficial muscles of the head and neck into the subclavian vein.

The **inferior vena cava,** which empties into the heart, is formed by the juncture of the left and right common iliac veins. It drains the lower extremities, the pelvis, and the abdominal region. The common iliac vein is formed by the junction of the external and internal iliac veins. The external iliac vein, which is a continuation of the femoral vein (discussed in chapter 13) drains the lower extremity. The internal iliac vein and its branches drain structures of the lower abdominal and pelvic regions.

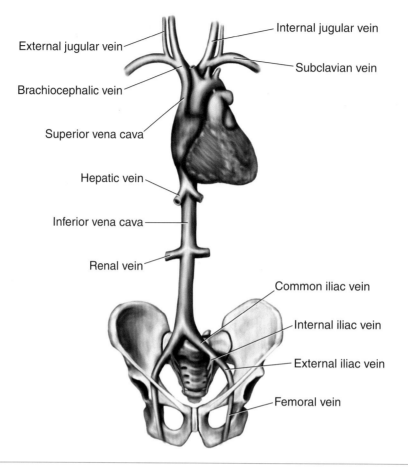

Figure 9.5 Major veins of the trunk, thorax, and abdomen.

REVIEW OF TERMINOLOGY

The following terms were discussed in this chapter. Define or describe each term, and where appropriate, identify the location of the named structure either on your body or in an appropriate illustration.

abdominal aorta
abdominal parietal branches (of thoracic aorta)
abdominal visceral branches (of thoracic aorta)
aortic arch
ascending aorta
basilar artery
brachiocephalic artery
brachiocephalic vein
cervical plexus
common carotid artery
common iliac artery
coronary artery
descending aorta

diaphragm
external carotid artery
external iliac artery
external jugular vein
femoral artery
inferior vena cava
intercostal nerves
internal carotid artery
internal iliac (hypogastric) artery
internal jugular vein
lumbar plexus
lumbosacral plexus
phrenic nerve
posterior cerebral arteries

pudendal plexus
sacral plexus
spinal accessory
subclavian artery
subcostal nerve
superior vena cava
thoracic aorta
thoracic nerves
thoracic parietal branches (of thoracic aorta)
thoracic visceral branches (of thoracic aorta)
vertebral arteries
vertebral vein

SUGGESTED LEARNING ACTIVITY

1. With your index and middle finger, locate your Adam's apple. Move laterally until your fingers encounter a band of muscle fibers. Apply gentle pressure with your fingertips.

   a. Do you feel a pulse?

   b. What vessel are you palpating to get that pulse?
   c. What muscle are you palpating to find this pulse?

MULTIPLE-CHOICE QUESTIONS

1. Which of the following structures does *not* belong in the same group as the other three structures?

   a ascending aorta
   b. descending aorta
   c. abdominal aorta
   d. thoracic aorta

2. Which of the following structures does *not* drain into the inferior vena cava?

   a. internal iliac vein
   b. jugular vein
   c. common iliac vein
   d. femoral vein

FILL-IN-THE-BLANK QUESTIONS

1. The diaphragm is innervated by the _____ _____ nerve.

2. The hypogastric artery is the _____ _____ branch of the _____ artery.

PART III   SUMMARY TABLE   Articulations of the Spinal Column, Pelvis, and Thorax

Joint	Type	Bones	Ligaments	Movement
Spinal column				
Atlanto-occipital	Condyloid	Occipital and atlas	• Articular capsules • Anterior atlanto-occipital • Posterior atlanto-occipital • Lateral atlanto-occipitals	Flexion, extension, abduction, adduction
Occipitoaxial	None	Axis and occipital	Tectoral, apical, and alar	None
Atlantoaxial	Pivot	Atlas and axis	• Articular capsules • Anterior atlantoaxial • Posterior atlantoaxial • Transverse	Rotation
Intervertebral	Amphiarthrodial (connected by cartilage)	Vertebral bodies	• Anterior longitudinal (occipital to sacrum) • Posterior longitudinal (occipital to coccyx) • Intervertebral discs	Slight gliding (limited flexion, extension, and rotation)
Intervertebral	Arthrodial (plane/gliding)	Vertebral articular processes and arches	• Articular capsules • Ligamentum flava (axis to sacrum) • Ligamentum nuchae (occipital to 7th cervical) • Supraspinous (thoracic to lumbar) • Interspinous (lumbar) • Intertransverse (lumbar)	Gliding
Thorax				
Costovertebral	Arthrodial (plane/gliding)	Vertebral bodies and heads of ribs	• Capsule • Costovertebral radiate • Interarticular	Gliding
Costotransverse	Arthrodial (plane/gliding)	Vertebral transverse processes and the necks and tubercles of ribs	• Capsule • Anterior costotransverse • Posterior costotransverse • Ligament of the neck • Ligament of the tubercle	Gliding
Sternocostal	Synarthrodial (fixed, joined by cartilage)	Manubrium of sternum and 1st rib	• Articular disc	None
Sternocostal	Arthrodial (plane/gliding)	Sternum and 2nd through 7th ribs	• Costosternal capsules • Costosternal radiates • Interarticulars • Costoxiphoid	Gliding
Costochondral	Synarthrodial (fixed)	Ribs and costocartilage	• Capsule • Periosteum	None
Interchondral	Synarthrodial (fixed)	6th, 7th, 8th, 9th ribs and costocartilages	• Capsule • Interchondral	None
Pelvis				
Sacroiliac	Amphiarthrodial (connected by cartilage)	Sacrum and ilium	• Anterior sacroiliac • (3) posterior sacroiliac (long, short, and interosseous) • Iliolumbar	Very limited (slight in pregnancy)

Pelvis				
Pubic symphysis	Amphiarthrodial (connected by cartilage)	Pubic bones	• Interpubic disc • Anterior pubic • Posterior pubic • Superior pubic • Inferior pubic (arcuate)	(Slight in pregnancy)
Nonarticular ligaments of the pelvis				
		Sacrum, coccyx, and ilium	• Sacrotuberous • Sacrospinous	(Stabilize the sacrum and coccyx)
		Anterior superior iliac spine to the pubic tubercle	Inguinal ligament	A landmark that separates the anterior abdominal wall from the thigh (the groin area)

PART III SUMMARY TABLE Spinal Column and Thorax Muscles, Nerves, and Blood Supply

Muscle	Origin	Insertion	Action	Nerve	Blood supply
Thorax					
Diaphragm	Sternum (xiphoid), last 7 ribs, first 4 lumbar vertebrae	Central tendon	Pulls central tendon down	Phrenic	Superior and inferior phrenic
Internal intercostals	Upper portion of costal groove	Upper edge of rib cartilage below	Elevation of rib	T2–T6	Intercostals
External intercostals	Lower edge of rib cartilage	Upper edge of rib below	Elevation of rib	T2–T6	Intercostals
Levatores costarum	Transverse process of C7 and T1–T11	Angle of rib below	Elevation of rib	T2–T6	Intercostals
Subcostals	At angle on inner aspect of rib	Inner aspect of 2nd and 3rd ribs below	Elevation of rib	T2–T6	Intercostals
Serratus posterior superior	Spines of C7 and T1–T3	Angles of 2nd, 3rd, 4th, and 5th ribs	Elevation of rib	1st–3rd intercostal	Intercostal and transverse cervical
Serratus posterior inferior	Spines of T11–T12 and L1–L3	Inferior edge of last 4 ribs	Depression of rib	10th and 11th intercostal	Intercostal and transverse cervical
Transversus thoracic	Distal 1/3 of posterior aspect of sternum and cartilage of 4th, 5th, 6th, and 7th ribs	Posterior aspect of 2nd, 3rd, 4th, 5th, and 6th rib cartilage	Elevation of rib	2nd–6th thoracic	Intercostals
Spinal column, cervical (anterior)					
Sternocleido-mastoid	Middle 1/3 of clavicle and sternum	Mastoid process	Flexion, lateral flexion, and rotation of cervical spine	Spinal accessory, 2nd and 3rd cervical	Transverse cervical
Prevertebrals					
Rectus capitis anterior	Lateral aspect of atlas	Occipital bone	Flexion and lateral flexion of cervical spine	C1 and C2	Pharyngeal and inferior thyroid
Rectus capitis lateralis	Transverse process of atlas	Inferior aspect of occipital bone	Flexion and lateral flexion of cervical spine	C1 and C2	Pharyngeal and occipital
Longus colli					
Vertical	Body of C5–C7 and T1–T3	Bodies of C2–C5	Flexion of cervical spine	C2–C8	Inferior and superior thyroid
Superior oblique	Anterior aspect of C3–C5 transverse processes	Anterior atlas	Flexion of cervical spine	C2–C8	Inferior and superior thyroid
Inferior oblique	T1–T3 bodies	C5–C6 transverse processes	Flexion of cervical spine	C2–C8	Inferior and superior thyroid

Scaleni					
Scalenus anterior	C3–C7 transverse processes	First rib	Lateral flexion of cervical spine and elevation of 1st rib	C4–C8	Transverse cervical and inferior thyroid
Scalenus medius	C2–C7 transverse processes	First rib	Lateral flexion of cervical spine and elevation of 1st rib	C4–C8	Cervical and inferior thyroid
Scalenus posterior	C4–C6 transverse processes	Second rib	Elevation of 2nd rib	C4–C8	Cervical and inferior thyroid
Spinal column, cervical (posterior)					
Rectus capitis posterior major	Spine of C2	Occipital bone	Extension, lateral flexion, and rotation of cervical spine	C1	Vertebral
Rectus capitis posterior minor	Posterior atlas	Occipital bone	Extension, lateral flexion, and rotation of cervical spine	C1	Vertebral
Obliquus capitis superior	Transverse process of atlas	Occipital bone	Extension of cervical spine	C1	Vertebral and occipital
Obliquus capitis inferior	Spine of the axis (C2)	Transverse process of the atlas (C1)	Rotation of cervical spine	C1	Vertebral
Splenius capitis	C7–T6 spinous processes	Mastoid process	Extension, lateral flexion, and rotation of cervical spine	C2–C4	Occipital and transverse cervical
Splenius cervicis	T4–T6 spinous processes	C1–C4 transverse processes	Extension, lateral flexion, and rotation of cervical spine	C2–C4	Occipital and transverse cervical
Spinal column, thoracic/lumbar (anterior)					
Rectus abdominis	Pubic symphysis	5th–7th rib cartilage and sternum	Trunk flexion and pelvic girdle flexion	Last 6 intercostals	Lower intercostals and epigastric
Obliquus internus	Lateral inguinal ligament, middle of iliac crest	Pubic bone to 7th, 8th, and 9th ribs	Trunk flexion and rotation	Last 3 intercostals, iliohypogastric, and ilioinguinal	Inferior epigastric branch of external iliac
Obliquus externus	Inferior edge of last 8 ribs	Outer edge of middle 1/2 of iliac crest	Trunk flexion, rotation, and pelvic girdle flexion	Last 7 intercostals, iliohypogastric	Lower intercostals and inferior epigastric branch of external iliac
Transversus abdominis	Inner aspect of last 6 ribs, internal aspect of middle 1/2 of iliac crest, lateral 1/3 of inguinal ligament	Pubic bone and linea alba	Flexion and rotation of trunk	Last 6 intercostals, iliohypogastric, ilioinguinal, and genitofemoral	Inferior epigastric branch of external iliac

(continued)

PART III SUMMARY TABLE *(continued)*

Muscle	Origin	Insertion	Action	Nerve	Blood supply
Spinal column, thoracic/lumbar (posterior)					
Spinals					
Spinalis capitis	C7 and T1–T7 transverse processes	Occipital bone	Extension of cervical spine	T6–T9	Occipital
Spinalis cervicis	T4–T6 spinous processes	C2, C3, and axis spinous processes	Extension of cervical spine	T6–T9	Vertebral and transverse cervical
Spinalis dorsi (thoracis)	T11, T12, L1, and L2 spinous processes	T4–T8 spinous processes	Extension of vertebral column	T6–T9	Intercostals
Semispinals					
Semispinalis capitis	C4–C7 articular processes and T1–T6 transverse processes	Occipital bone	Extension, lateral flexion, and rotation of cervical spine	C1–C5	Occipital, transverse cervical, vertebral, and deep cervical
Semispinalis cervicis	T1–T6 transverse processes	C2–C5 spinous processes	Extension and lateral flexion of cervical spine	T3–T6	Deep cervical, occipital, and vertebral
Semispinalis dorsi (thoracis)	T6–T10 transverse processes	C6–C7 and T1–T4 spinous processes	Extension of cervical spine	T3–T6	Intercostals
Iliocostals					
Iliocostalis cervicis	Angle of 3rd, 4th, 5th, and 6th ribs	C4–C6 transverse processes	Lateral flexion and rotation of cervical spine	Cervical	Thyrocervical
Iliocostalis dorsi (thoracis)	Angle of last 6 ribs	Upper edge of first 6 ribs and C7 transverse process	Lateral flexion and rotation of spine, depression of ribs	Thoracic	Intercostals
Iliocostalis lumborum	Posterior aspect of iliac crest	Angle of last 7 ribs	Extension and rotation of spine	Lumbar	Lumbar

Longissimi				
Longissimus capitis	C4–C7 articular processes and T1–T5 transverse processes	Mastoid process	Extension, lateral flexion, and rotation of cervical spine	C1–L5	Transverse cervical, external carotid, and vertebral spine
Longissimus cervicis	T1–T5 transverse processes	C2–C6 transverse processes	Extension, lateral flexion, and rotation of cervical spine	C1–L5	Transverse cervical, external carotid, and vertebral spine
Longissimus dorsi (thoracis)	L1–L5 transverse processes and posterior iliac crest	3rd–12th ribs and T1–T12 transverse processes	Extension, lateral flexion, and rotation of cervical spine; rib depression	C1–L5	Intercostals and lumbar
Sacrospinalis (erector spinalis)	Lumbar and sacral spinous processes	See: iliocostalis lumborum, longissimus dorsi, and spinalis dorsi	Extension, lateral flexion, and rotation of spine	Thoracic, lumbar, sacral	Lumbar and intercostal
Quadratus lumborum	Lower lumbar transverse processes, posterior aspect of iliac crest	Lowest edge of 12th rib and L1–L4 transverse processes	Extension and lateral flexion of spine, depression of 12th rib	L1–L4	Lumbar and lower intercostals
Multifidus	C4–C7 articular processes, T1–T12 and L1–L5 transverse processes, posterior superior iliac spine, and posterior sacrum	S5–C4 spinous processes	Extension and rotation of spine	C4–S5	Lumbar

169

PART IV

Lower Extremity

The Hip and Thigh

Any discussion of the hip must include the bones of the pelvis and thigh and the muscles of the thigh as well as the ligaments and muscles of the hip. Several muscles that cross the hip joint also cross the knee joint, and discussion of these muscles in this chapter is restricted to their function at the hip joint. Their function at the knee joint is discussed in the next chapter.

Bones of the Hip Joint and Thigh

The hip joint is the articulation between the pelvis and the **femur**, or thigh bone. The pelvis is discussed in chapter 7. The important structure of the pelvis regarding the hip joint is the anatomical area labeled the **acetabulum** (figure 10.1; also see figure 7.32 on page 136). Remember that the acetabulum is formed by the articulation of the

three bones making up the **innominate bone** (the **ilium,** the **ischium,** and the **pubic bone**). The hip joint is classified as a ball-and-socket or triaxial joint. The acetabulum is considered the socket of the joint, and the ball of the hip joint is the structure known as the **head** of the femur (figure 10.2). Although the hip joint and the shoulder joint are often compared because of their similarities as ball-and-socket or triaxial joints, the hip joint is a much more stable joint than the shoulder because of the depth of the acetabulum compared with the very shallow **glenoid fossa** of the scapula. It is through the hip joint that the entire weight of the trunk and upper extremities is transferred to the lower extremities.

The femur, the longest and largest bone in the body, is discussed in this chapter only as it pertains to the hip joint (at the proximal end). Further discussion of the distal end of the femur is presented in chapter 11 on the knee.

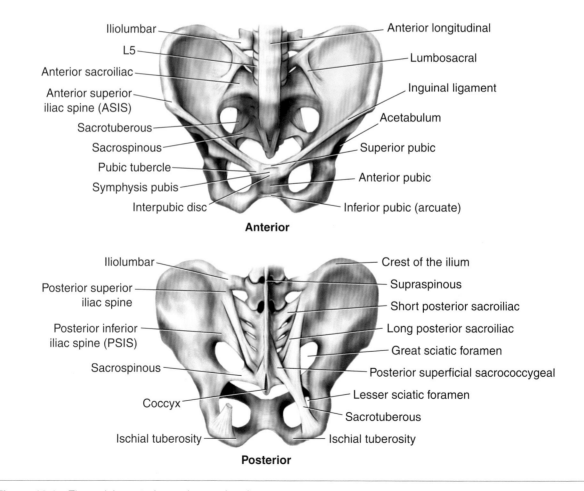

Figure 10.1 The pelvis, anterior and posterior views.

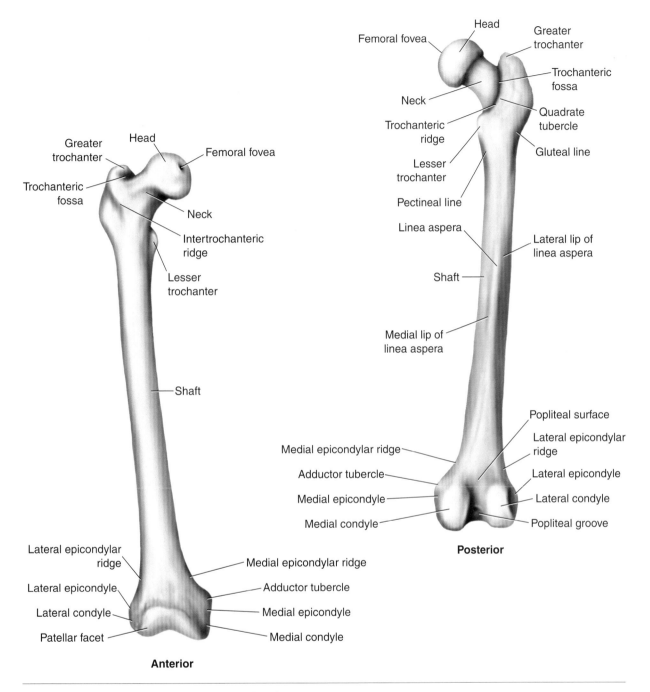

Figure 10.2 The femur, anterior and posterior views.

The ball, or head, of the femur is separated from the shaft of the bone by the **neck** of the femur. At the proximal end of the shaft of the femur, just distal to the neck, are two large projections of bone known as the **greater** and **lesser trochanters.** These structures serve as points of insertion for the major muscles responsible for hip joint actions.

Hands on... Rotate one leg toward the other leg (internally rotate your hip joint) and feel the greater trochanter as it moves against your skin (figure 10.3).

The lesser trochanter is distal, medial, and somewhat posterior to the greater trochanter and more readily observed in a posterior view of the

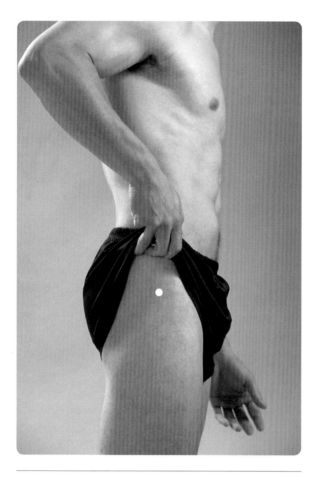

Figure 10.3   Locating the greater trochanter of the femur.

femur. The ridge of bone running between the lesser and greater trochanters is known as the **intertrochanteric ridge** (see figure 10.2). A large ridge or crest of bone running down the posterior aspect of the shaft of the femur is known as the **linea aspera,** Latin for "rough line" (see figure 10.2). The linea aspera serves as a source of muscular attachment for numerous muscles of both the hip and knee joints.

Ligaments of the Hip Joint

Although the hip joint (figure 10.4) derives most of its stability from the bony formation between the head of the femur and the acetabulum of the pelvis, it is also reinforced by several strong ligaments. There are seven ligaments of the hip joint. As with all synovial joints, there is a **capsular ligament** running from the edge of the acetabulum of the pelvis to the neck of the femur. The **glenoid**

lip (acetabular labrum), as in the shoulder joint, is a fibrocartilaginous rim on the outer edge of the acetabulum that helps to deepen its socket. Three ligaments that strengthen the capsular ligament are easily identified because their names represent the bones they tie together: the **iliofemoral,** the **ischiofemoral,** and the **pubofemoral ligaments** (figures 10.5 and 10.6). The iliofemoral ligament looks like an upside-down Y and therefore is frequently referred to as the Y-ligament. Because of the anatomical relationships between the ilium and the femur and between the pubic bone and the femur, the iliofemoral and pubofemoral ligaments cross the anterior aspect of the hip joint and attach to the neck of the femur. Likewise, because the ischium is posterior to the femur, the ischiofemoral ligament crosses the posterior aspect of the hip joint and attaches to the neck of the femur. The **transverse acetabular ligament** spans a small notch in the lower portion of the glenoid lip and creates a foramen that blood vessels pass through to and from the hip joint (figure 10.7). The last ligament, the **ligamentum capitis femoris** (figure 10.7), runs between the transverse acetabular ligament and a hole in the superior edge of the femoral head known as the **femoral fovea** (see figure 10.2). This ligament attaches the head of the femur to the acetabulum and plays the least significant role of any of the ligaments of the hip joint.

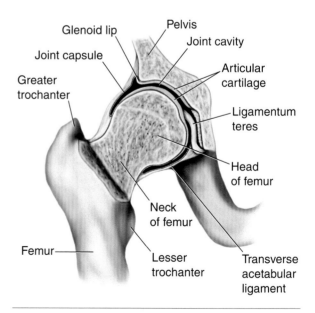

Figure 10.4   A longitudinal section of the hip joint.

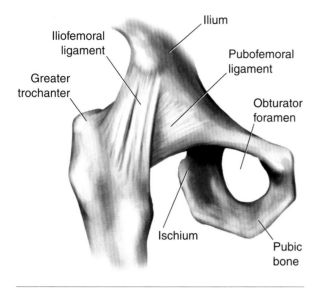

Figure 10.5 The iliofemoral (or Y) and pubofemoral ligaments, anterior view.

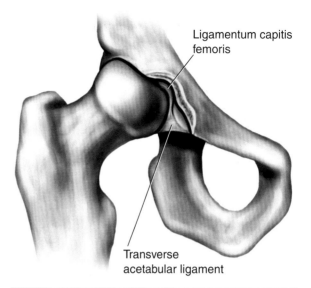

Figure 10.7 The transverse acetabular ligament and ligamentum capitis femoris, anterior view.

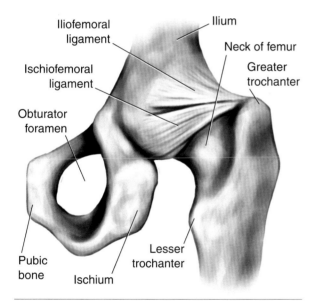

Figure 10.6 The ischiofemoral ligament, posterior view.

Fundamental Movements of the Hip Joint

The hip joint is classified as a **triaxial joint,** having movement in all three planes about all three axes: flexion and extension in the sagittal plane about a frontal horizontal axis, abduction and adduction in the frontal plane about a sagittal-horizontal axis, and internal and external rotation in the horizontal plane about a vertical axis. The musculature that creates those six fundamental movements includes flexors, extensors, abductors, adductors, and rotators. They are presented in groups based on their position: anterior, posterior, lateral, or medial to the hip joint.

Hands on . . . As you stand in the anatomical position, flex your hip. What group of muscles primarily flexed your hip (anterior, posterior, lateral, medial)? Return to the anatomical position and extend your hip joint. Again, what muscle group (anterior, posterior, lateral, medial) was primarily involved in extending your hip? Now abduct and then adduct your hip joint. Determine which muscle groups are primarily responsible for these actions of the hip joint. To determine which muscle groups are primarily responsible for internal and external rotation of the hip joint, lie on your back and internally and externally rotate your hip joint while palpating the muscles surrounding your hip joint. Now that you have completed the hip exercises, we can look at the muscles that cross the hip joint.

Muscles of the Hip Joint and Upper Leg

Muscles that cross the hip joint are grouped by position as anterior, posterior, medial, or lateral to the joint.

Anterior Muscles

From the previous exercise, you know that the muscles that performed flexion of your hip are located anterior to the hip joint. Interestingly enough, some of these muscles also cross the knee joint. In this chapter, however, only their function at the hip joint is considered. The anterior muscles are listed here (figure 10.8).

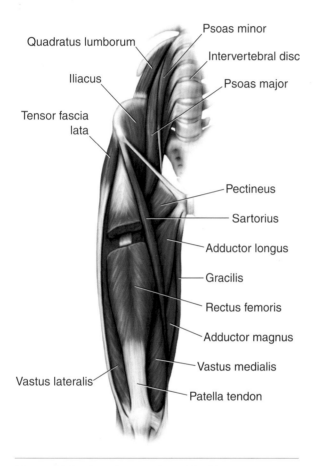

Figure 10.8   Anterior muscles of the hip.

> **Iliopsoas:** The iliopsoas muscle is a combination of the iliacus and psoas major muscles. These muscles, along with the psoas minor, are commonly referred to as the *true groin muscles* (or *hip flexor muscle group)*: The term *groin* refers to the anterior aspect of the hip joint. Their main function is to flex the hip joint, but if the femur is stabilized, these muscles cause flexion of the lumbar spine when they contract. The medial muscles of the hip joint (the adductors) are often referred to as the *common groin muscles* (or hip adductor muscle group) and are discussed later in this chapter.

> **Psoas major:** This muscle originates on the transverse processes of all five **lumbar vertebrae** and the bodies and **intervertebral discs** of the 12th **thoracic vertebra** and all five lumbar vertebrae. The psoas major inserts on the lesser trochanter of the femur. Along with the iliacus muscle, this muscle flexes the hip joint, assists with adduction and external rotation of the hip joint, and assists with flexion and rotation of the lumbar spine (figure 10.9).

> **Iliacus:** This muscle originates from the iliac fossa of the pelvis to the tendon of the psoas major and inserts just distal to the lesser trochanter of the femur. Along with the psoas major muscle, this muscle flexes the hip joint, assists with adduction and external rotation of the hip joint, and, when the femur is stabilized, assists with flexion of the lumbar spine (figure 10.9).

> **Psoas minor:** This muscle is not present on one or both sides in more than 50% of individuals. The muscle originates from the 12th thoracic and 1st lumbar vertebrae and the intervertebral disc between these two vertebrae and inserts on the pubic bone. Therefore, the psoas minor is *not* considered a muscle of the hip joint. Its function is to assist the psoas major with flexion of the lumbar spine (figure 10.9).

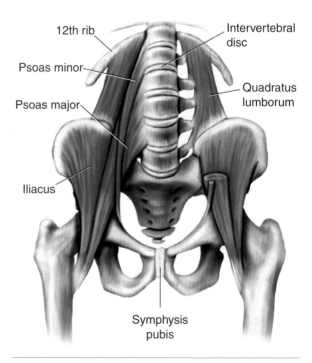

Figure 10.9   The psoas major, psoas minor, and iliacus.

> **Sartorius:** This muscle, the longest muscle of the body, originates from the anterior superior spine of the ilium and inserts on the medial aspect of the proximal end of the tibia, a bone of the lower leg (see chapter 11, figure 11.5), just inferior to the medial condyle (figure 10.8). Note that this muscle crosses both the hip and knee joints. The functions of the sartorius at the hip joint are flexion, abduction, and external rotation.

Hands on . . . Sitting on the floor with your ankles crossed and your knees apart (the tailor's position), you should be able to palpate your sartorius just beneath the anterior superior iliac spine (figure 10.10). The terms *sartorius* (from a Latin word for tailor) and *tailor's position* both refer to tailors' practice in bygone days of sitting for hours to do their work in a cross-legged position, which often caused soreness in this muscle after a long day of sewing.

> **Rectus femoris:** This muscle is part of a powerful group of four muscles known as the **quadriceps femoris.** Three of these muscles, the vastus lateralis, the vastus intermedius, and the vastus medialis, cross only the knee joint, but the rectus femoris crosses both the knee and hip joints. The muscle originates on the anterior inferior iliac spine and inserts on the tibial tuberosity. The rectus femoris flexes the hip joint and also acts as a very weak abductor.

Hands on . . . Lie on a table with your hips in extension and your knees flexed over the end of the table, and extend your knee against a partner's manual resistance. Note the prominence of muscular tissue running the length of the anterior thigh area (figure 10.11). This prominence is the rectus femoris contracting against your resistance.

> **Tensor fascia lata:** This muscle originates on the external rim of the crest of the ilium and combines with the gluteus maximus muscle to form another structure, the **iliotibial band** or **tract,** which inserts in the area of the lateral condyle of

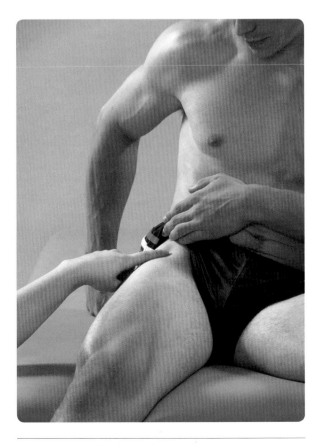

Figure 10.10   Finding the sartorius.

Figure 10.11   Locating the rectus femoris.

the tibia (figures 10.8, 10.12, and 11.5). The tensor fascia lata flexes and abducts the hip joint. Even though the term *lata* might lead one to believe that this is a lateral muscle of the hip joint, it is actually considered an anterior hip joint muscle, although a very lateral anterior muscle. More about this muscle and its involvement with the iliotibial band is presented later in this chapter and in chapter 11 on the knee joint.

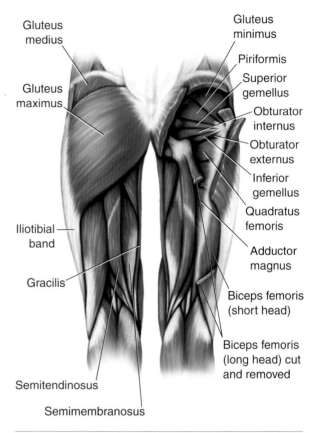

Figure 10.13   Posterior muscles of the hip.

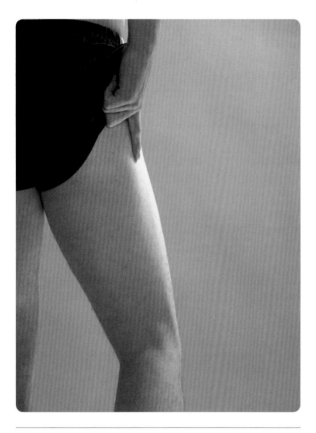

Figure 10.12   Demonstrating the tensor fascia lata and iliotibial band.

> **Pectineus:** This muscle originates between the anterior superior aspect of the pubic bone to the femur just distal to the lesser trochanter (figure 10.8). Besides crossing the hip anteriorly, making it a hip flexor, it is also a strong adductor and external rotator of the hip because of its angle of pull.

Posterior Muscles

Ten posterior muscles cross the hip joint (figure 10.13), including a group of three (the **hamstrings**)

and a group commonly referred to as the six **deep external rotators.** The hamstrings are discussed only as they function at the hip joint. They also cross the knee joint and are discussed again in chapter 11.

> **Biceps femoris:** Although all three hamstring muscles (biceps femoris, semitendinosus, and semimembranosus) originate on the **ischial tuberosity** and have the same functions—extension of the hip joint and flexion of the knee joint—the biceps femoris is unique in that it has two heads: The **long head** originates on the ischial tuberosity with the semitendinosus and semimembranosus, and the **short head** originates from the linea aspera of the femur. Both heads combine into the belly of the muscle, which then inserts on the head of the **fibula,** a bone of the lower leg (see chapters 11 and 12), and the lateral condyle of the tibia (figure 11.5). The biceps femoris functions specifically at the hip joint to extend the joint and assist in adduction and external rotation of the joint.

Hands on . . . The tendon of insertion of the biceps femoris can be observed on the posterior lateral aspect of the knee joint when you lie in the position illustrated in figure 10.14.

> **Semitendinosus:** Another muscle of the hamstrings group, the semitendinosus also originates on the ischial tuberosity and inserts on the medial aspect of the tibia just below the medial condyle (figure 11.5).

Hands on . . . The tendon of insertion of the semitendinosus has a very prominent, cordlike structure (figure 10.15). Have your partner lie in a prone (face down) position and flex his or her knee joint, and then locate this tendon on the posterior medial aspect of the knee joint. The muscle extends and assists with internal rotation and adduction of the hip joint.

> **Semimembranosus:** The final muscle of the hamstrings group is the semimembranosus, which, like the semitendinosus, originates on the ischial tuberosity and inserts on the medial condyle of the tibia (figure 11.5). This muscle extends the hip joint and assists with internal rotation and adduction of the hip joint.

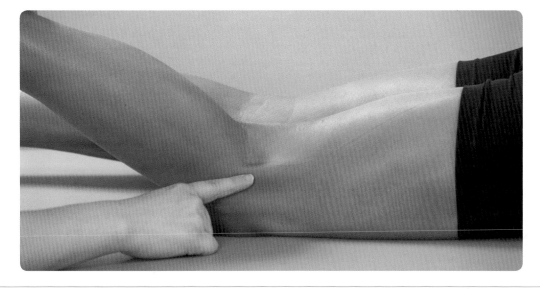

Figure 10.14   Locating the biceps femoris.

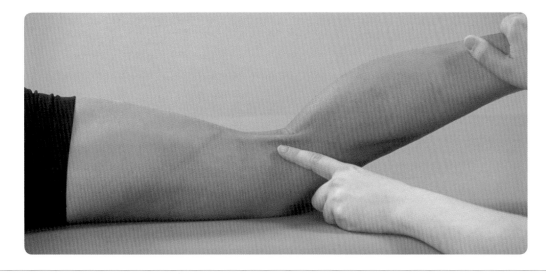

Figure 10.15   Finding the semitendinosus.

Hands on . . . The semimembranosus tendon of insertion is a very broad structure just medial to the semitendinosus tendon of insertion (figure 10.16). Ask your partner to assume the same position as when you located the semitendinosus, and then locate the semimembranosus.

> **Gluteus maximus:** This rather large muscle is often identified as the buttocks muscle. It originates from the posterior aspects of the ilium, **sacrum,** and **coccyx** bones and inserts by blending with the tensor fascia lata muscle to form the iliotibial band (figure 10.13). The muscle extends and externally rotates the hip joint. Because one portion of the belly of the muscle crosses above the hip joint and another portion below the hip joint, the muscle is involved in both abduction and adduction of the hip. Look carefully at the hip joint in figure 10.13, and determine which portion of the gluteus maximus assists with abduction and which portion assists with adduction of the hip joint.

Hands on . . . Deep below the gluteal muscles is a group of six muscles that perform external rotation of the hip joint. Lying on your back, legs straight, have your partner stabilize one of your feet in the anatomically neutral position. Attempt to move your foot away from the other foot by externally rotating the entire lower extremity. Apply hand pressure into the belly of the gluteus maximus muscle of the rotating leg. Did you feel contraction of muscles deep within the joint? What muscles were contracting? Let's look for

Figure 10.16   Locating the semimembranosus.

Focus on . . . HAMSTRING STRAIN

The term *hamstring strain* is often used and often misunderstood. There are three muscles, biceps femoris, semitendinosus, and semimembranosus, that could be strained as a result of hip hyperflexion or knee hyperextension or, more likely, a combination of both hip flexion and knee extension. Although determining which of the three muscles (or combination of muscles) is strained is of little importance to the general public, someone with a good knowledge of human anatomy should be able to understand that *hamstring strain* is a somewhat nonspecific term. Flexibility is a major factor in preventing hamstring strain, because many athletic activities may require the hip joint to flex at the same time the knee joint is extending, which puts this muscle group under tension. (Picture a hurdler's lead leg as it goes over a hurdle.)

these answers in the discussion of the six deep external rotator muscles of the hip joint (figure 10.13), which are presented anatomically from superior to inferior.

> **Piriformis:** This muscle originates on the upper portion of the sacrum, crosses the **greater sciatic notch,** and inserts on the upper surface of the greater trochanter. If contracted in spasm (sudden, involuntary contraction) as the result of strain or physical trauma, this muscle can impinge the sciatic nerve as it passes through the **sciatic notch** of the pelvis, causing pain to radiate down the lower extremity along the course of the sciatic nerve and its branches.

> **Superior gemellus:** Originating from the **spine of the ischium** to the medial aspect of and inserting on the greater trochanter, the superior gemellus externally rotates the hip joint.

> **Internal obturator (obturator internus):** Originating from the inner surface of the **obturator foramen** of the pelvis and inserting on the medial aspect of the greater trochanter, the internal obturator externally rotates the hip joint.

> **Inferior gemellus:** Originating from the ischial tuberosity and inserting on the medial aspect of the greater trochanter, the inferior gemellus externally rotates the hip joint.

> **External obturator (obturator externus):** Originating from the outer surface of the pubic bone and ischium around the obturator foramen and inserting on the pelvis to the **trochanteric fossa** of the femur, the external obturator externally rotates the hip joint.

> **Quadratus femoris:** The quadratus femoris originates on the ischial tuberosity, inserts on the intertrochanteric ridge of the femur, and externally rotates the hip joint.

Medial Muscles

Three of the four adductor muscles of the hip joint give their function away by their names. The fourth adductor, the gracilis, is the only one of the four that also crosses the knee joint. These four muscles (figure 10.17), as mentioned earlier, are known as the *common groin muscles (or adductor group muscles)*. Compared with the *true groin* muscles *(the iliopsoas muscle group)*, these groin muscles

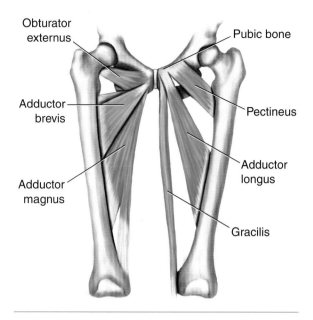

Figure 10.17   Adductors of the hip.

are more commonly affected when one experiences a groin strain.

> **Adductor longus:** This muscle originates from the body of the pubic bone and inserts on the middle third of the linea aspera of the femur. Its major function is adduction of the hip, and it assists with flexion and external rotation.

> **Adductor brevis:** This muscle originates from the body and inferior surface of the pubic bone and inserts on the proximal third of the linea aspera of the femur. Its major function is adduction of the hip, and it assists with external rotation and flexion.

> **Adductor magnus:** This muscle (figures 10.17 and 10.18) originates on the inferior surface of the pubic bone, the ischium, and the ischial tuberosity and inserts on the entire length of the linea aspera and the **adductor tubercle** of the femur (see chapter 11). The adductor magnus has two distinct portions: the anterior portion, which adducts, flexes, and externally rotates the hip joint, and the posterior portion, which adducts, extends, and internally rotates the hip joint.

> **Gracilis:** This muscle originates on the lower portions of the anterior aspect of the **pubic symphysis** and the pubic bone and inserts on the tibia just distal to the medial condyle. At the hip joint, the gracilis functions to adduct and flex the joint.

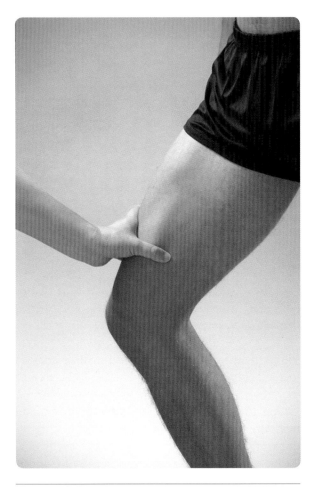

Figure 10.18   Locating the adductor magnus.

Sitting in the lotus, or butterfly, position (sit on the floor, place the soles of your feet together with your knees flexed and your hips externally rotated)), you can easily palpate your adductor longus and gracilis muscles in the groin area. The gracilis is the more lateral of the two tendons. Additionally, you can palpate the gracilis by pressing with your finger on the area just medial to the semimembranosus tendon of insertion on the medial aspect of the knee joint (figure 10.19).

Lateral Muscles

There are only two lateral muscles of the hip joint, not including the tensor fascia lata, which was identified as an anterior, not lateral, muscle of the hip joint. Even though the gluteus maximus shares a name with the lateral muscles of the hip joint, it is a posterior muscle. The lateral muscles of the hip joint are the gluteus medius and the gluteus minimus (figure 10.13).

> **Gluteus medius:** The gluteus medius originates from the middle of the external aspect of the ilium to the iliac crest and inserts on the posterior lateral aspect of the greater trochanter (figures 10.13 and 10.20). The major function of this muscle is hip abduction. The anterior por-

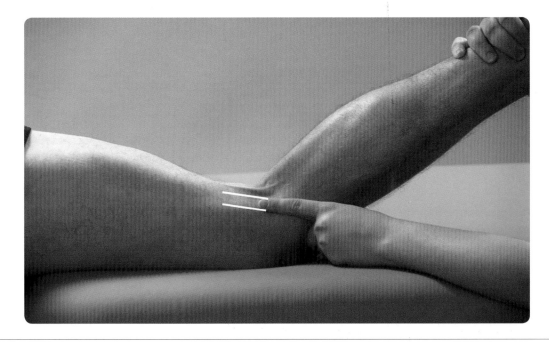

Figure 10.19   Identifying the gracilis.

tion assists with flexion and internal rotation, and the posterior portion assists with extension and external rotation.

> **Gluteus minimus:** Originating in the same area on the ilium as the gluteus medius, the gluteus minimus inserts on the anterior surface of the greater trochanter. The major functions of this muscle are hip abduction and internal rotation. The anterior portion assists with flexion, and the posterior portion assists with extension.

Iliotibial Band

Iliotibial band refers to two muscles that cross the hip joint: the tensor fascia lata and the gluteus maximus (figure 10.13). The iliotibial band passes back and forth across the greater trochanter of the femur as the hip joint flexes and extends. To facilitate this movement of the iliotibial band over the greater trochanter, a large trochanteric bursa is located between the iliotibial band and the greater trochanter. Excessive flexion or extension or direct trauma to this area can cause inflammation of the bursa (trochanteric bursitis).

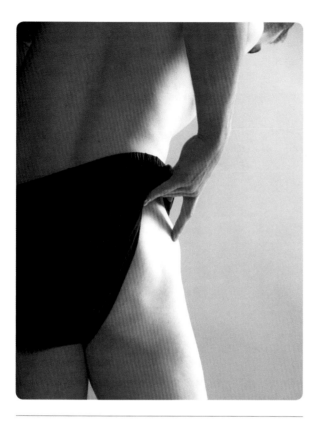

Figure 10.20   Locating the gluteus medius.

Focus on . . . GROIN STRAIN

Often we hear about athletes suffering from a strained groin injury. What muscle group is being referenced when the term **strained groin** is mentioned? This chapter has presented two muscle groups as being "groin muscles"; the iliopsoas muscle group (iliacus and psoas major muscles), often referred to as the *true groin muscles,* and the adductor muscle group (adductor longus, adductor brevis, adductor magnus, and the gracilis muscles), referred to as the *common groin muscles.* The true groin muscle (the iliopsoas group) is typically strained as the result of hyperextension of the hip joint. The common groin strain (the adductor group) is usually the result of overstretching of this muscle group. As mentioned earlier in this chapter, the adductor group is more commonly the muscle group that is strained in athletic endeavors and is therefore referred to as the common groin strain.

REVIEW OF TERMINOLOGY

The following terms were discussed in this chapter. Define or describe each term, and where appropriate, identify the location of the named structure either on your body or in an appropriate illustration.

acetabulum	adductor tubercle	external obturator
adductor brevis	biceps femoris	femoral fovea
adductor longus	capsular ligament	femur
adductor magnus (anterior and posterior portions)	coccyx	fibula
	deep external rotators	glenoid fossa

(continued)

REVIEW OF TERMINOLOGY   *(continued)*

glenoid lip
gluteus maximus
gluteus medius
gluteus minimus
gracilis
greater sciatic notch
greater trochanter
hamstrings
head of the femur
iliacus
iliofemoral ligament
iliopsoas
iliotibial band (tract)
ilium
inferior gemellus
innominate bone
internal obturator
intertrochanteric ridge

intervertebral disc
ischial tuberosity
ischiofemoral ligament
ischium
lesser trochanter
ligamentum capitis femoris
linea aspera
long head of the biceps femoris
lumbar vertebrae
neck of the femur
obturator foramen
pectineus
piriformis
psoas major
psoas minor
pubic bone
pubic symphysis
pubofemoral ligament

quadratus femoris
quadriceps femoris
rectus femoris
sacrum
sartorius
sciatic notch
semimembranosus
semitendinosus
short head of the biceps femoris
spine of the ischium
strained groin
superior gemellus
tensor fascia lata
thoracic vertebrae
transverse acetabular ligament
triaxial joint
trochanteric fossa

SUGGESTED LEARNING ACTIVITIES

1. Take the position of a runner in the starting blocks for a short running race. Place one foot forward and one foot back while placing your hands on the starting line.

   a. In what position is the hip joint of the forward leg (flexed, extended)?

   b. In what position is the hip joint of the back leg (flexed, extended)?

   c. Now imagine the starter has fired the starting gun. Into what position did the hip joint of the forward leg move, and what muscles caused that movement?

   d. Into what position did the hip joint of the back leg move, and what muscles caused that movement?

2. Perform the classic jumping jack exercise several times, observing the movements of the hip joint during the counts of 1 and 2.

   a. Describe the position of both hip joints on the count of 1 and the muscles that were used to put the hips in that position.

   b. Describe the position of both hip joints on the count of 2 and the muscles that were used to put the hips in that position.

3. Take one step forward with your right leg. Cross your left leg over your right leg so that your left foot is perpendicular to your right foot. Your left heel should now be near the outer edge of your right foot.

   a. Describe the position of your left hip.

   b. Describe the position of your right hip.

MULTIPLE-CHOICE QUESTIONS

1. Which of the following bones of the pelvis is not part of the structure known as the acetabulum?

   a. ilium
   b. ischium
   c. sacrum
   d. pubic bone

2. The greater trochanter of the femur is located where in relation to the hip joint?

   a. lateral
   b. medial
   c. anterior
   d. posterior

3. Which of the following is not considered a fundamental movement of the hip joint?

   a. flexion
   b. circumduction
   c. extension
   d. adduction

4. Which of the following ligaments is not considered a ligament of the hip joint?

   a. iliofemoral
   b. pubofemoral
   c. ischiofemoral
   d. sacroiliac

5. Which of the following muscles is absent on one or both sides in approximately 50% of human beings?
   a. psoas major
   b. psoas minor
   c. iliacus
   d. iliopsoas

6. Which of the following muscles is not part of the muscle group known as the iliopsoas?
   a. iliacus
   b. sacroiliac
   c. psoas minor
   d. psoas major

7. Which of the following muscles is not considered an anterior muscle of the hip joint?
   a. rectus femoris
   b. biceps femoris
   c. sartorius
   d. tensor fascia lata

8. Which of the following muscles is not an adductor of the hip joint?
   a. gracilis
   b. gluteus maximus
   c. pectineus
   d. gluteus medius

9. Which of the following muscles of the hip joint does not attach to the pubic bone?
   a. pectineus
   b. adductor brevis
   c. gracilis
   d. iliacus

10. Which of the following muscles is not considered one of the six deep external rotators of the hip joint?
    a. quadratus femoris
    b. rectus femoris
    c. piriformis
    d. internal obturator

FILL-IN-THE-BLANK QUESTIONS

1. A joint defined as a triaxial, ball-and-socket joint of the lower extremity is the _____.

2. The ligament of the hip often referred to as the Y-ligament is the _____ ligament.

3. The iliotibial tract consists of the combined tendons from the tensor fascia lata and the _____ _____.

4. The large tuberosity on which the hamstrings originate is part of the _____.

5. The psoas major and psoas minor muscles originate on the _____.

6. The six deep external rotators of the hip joint are located on the _____ aspect of the joint.

The Knee

The knee joint, one of the largest joints in the body, is a uniaxial, synovial joint and is often referred to as a hinge joint. In reality, the knee joint is *not* a true hinge joint but rather a modified hinge joint. A true hinge joint, like a door hinge, opens and closes about a single constant axis. In the knee joint, the tibia (distal bone) glides around the distal end of the femur (proximal bone), and although the movement remains in one plane (the sagittal plane), it occurs about an ever-changing axis. With each degree of movement in the sagittal plane, the frontal horizontal axis changes. Hence, the term *modified hinge joint* is more appropriate than just *hinge joint* for the knee joint. Although the knee joint appears to be well built structurally, it was not built to withstand many of the stresses placed on it by athletic activities. As we examine the normal anatomy of the joint, we also draw attention to the common results of abnormal stresses.

Bones of the Knee

The **femur** (figure 11.1) was presented in chapter 10 on the hip joint, but its role at the knee was left for discussion in this chapter. At the distal end of the shaft of the femur, the bone flares out, forming a **medial** and **lateral epicondylar ridge**, similar to those in the humerus just proximal to

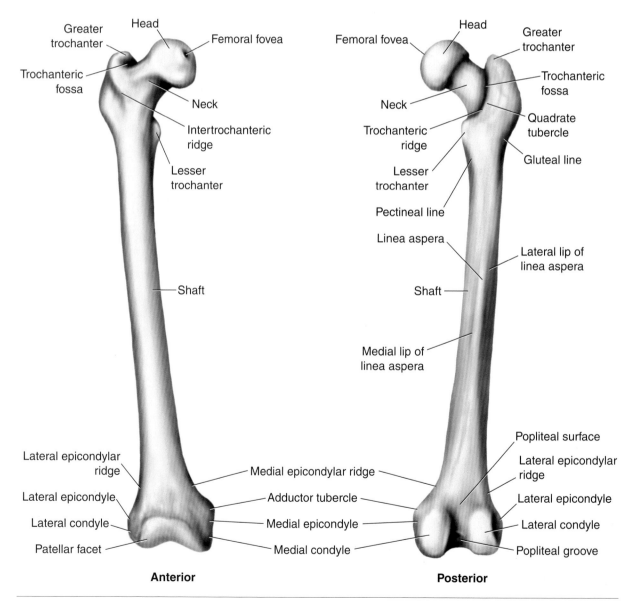

Anterior — Greater trochanter, Head, Femoral fovea, Trochanteric fossa, Neck, Intertrochanteric ridge, Lesser trochanter, Shaft, Lateral epicondylar ridge, Lateral epicondyle, Lateral condyle, Patellar facet, Medial epicondylar ridge, Adductor tubercle, Medial epicondyle, Medial condyle

Posterior — Femoral fovea, Head, Greater trochanter, Trochanteric fossa, Neck, Quadrate tubercle, Trochanteric ridge, Gluteal line, Lesser trochanter, Pectineal line, Linea aspera, Lateral lip of linea aspera, Shaft, Medial lip of linea aspera, Popliteal surface, Lateral epicondylar ridge, Lateral epicondyle, Lateral condyle, Popliteal groove

Figure 11.1 The femur, anterior and posterior views.

the elbow joint. Just distal to these ridges are the **medial** and **lateral epicondyles** (figure 11.2). Two other large prominences of bone, easily palpated just proximal to the joint line (an imaginary line between the femur and the tibia), are the **medial** and **lateral condyles** (figure 11.3). It is these two condyles that articulate with the tibia to transfer the body weight from the femur to the lower leg. The medial condyle is slightly more distal than the lateral condyle.

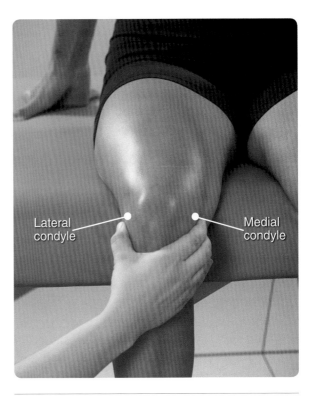

Figure 11.3 Locating the medial and lateral femoral condyles.

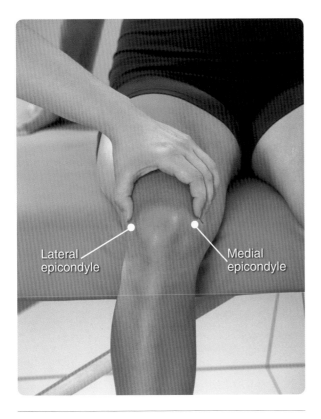

Figure 11.2 Finding the medial and lateral femoral epicondyles.

Hands on . . . After locating the lateral and medial femoral condyles (figure 11.3), move your fingers distal (below) the condyles. You should feel a space between the distal ends of the femur and the proximal ends of the tibia. This space is commonly known as the *joint line*. Under your fingers lying on the lateral and medial joint line lie the structures known as the *anterior horns* of the lateral and medial meniscus.

Just proximal to the medial epicondyle is a small prominence known as the **adductor tubercle** (figure 11.4). On the anterior surface of the

distal end of the femur, between the lateral and medial condyles, is a smooth surface covered with articular cartilage known as the **patellar facet.** This surface articulates with the posterior aspect of the patella (kneecap, discussed later), forming what is known as the **patellofemoral joint.** On the

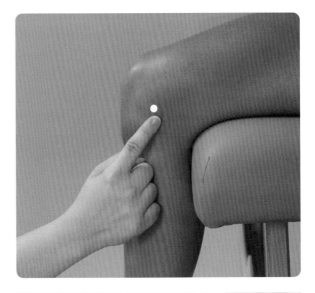

Figure 11.4 Identifying the adductor tubercle.

posterior side of the distal end of the femur, between the lateral and medial condyles, is a groove known as the **popliteal groove.** The area just proximal to this groove is known as the **popliteal surface.** The word *popliteal* refers to the posterior knee joint and is applied to bones, ligaments, muscles, nerves, or blood vessels of the posterior knee. The space between the femoral condyles is referred to as the **intercondylar notch** or the **femoral notch.**

The two bones of the lower leg are the **fibula** and **tibia** (figure 11.5). In this chapter on the knee joint, only the proximal ends of these two bones are described. Their distal ends are considered in chapter 12.

The fibula, the smaller, lateral bone, is essentially a non-weight-bearing bone that serves as a major source of soft tissue attachment at both the proximal and distal ends. Three prominent structures at the proximal end of the fibula are the **apex,** the **head,** and the **neck.** The tibia, the larger and medial bone of the lower leg, articulates with the

femur and bears the weight of the body from the femur to the foot. At the very proximal end of the shaft of the tibia are the **lateral** and **medial condyles,** and between them, on the proximal anterior surface of the tibia, lies a very large prominence known as the **tibial tuberosity.**

Hands on . . . You can easily palpate the tibial tuberosity by feeling for the large bump on the anterior surface of your lower leg just below your knee (figure 11.6).

A smaller prominence, located on the anterior aspect of the lateral condyle of the tibia, is known as **Gerdy's tubercle** (figure 11.7), where the iliotibial band inserts. On the posterior aspect of the proximal tibia, just distal to the space between the lateral and medial condyles, is the area known as the **popliteal surface.** Just beneath this surface is a small diagonal line known as the **popliteal line,** also known as the linea solei (see figure 11.5). The superior view of the tibia reveals several structures. The two large surfaces, the **medial** and **lateral condylar surfaces,**

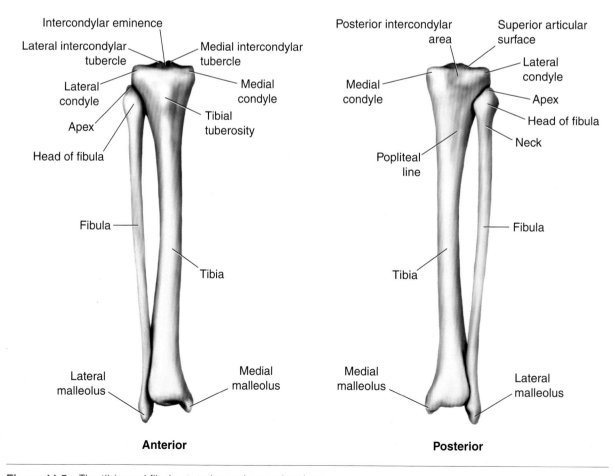

Labels (Anterior, left): Intercondylar eminence · Lateral intercondylar tubercle · Medial intercondylar tubercle · Lateral condyle · Medial condyle · Apex · Tibial tuberosity · Head of fibula · Fibula · Tibia · Lateral malleolus · Medial malleolus · **Anterior**

Labels (Posterior, right): Posterior intercondylar area · Superior articular surface · Medial condyle · Lateral condyle · Apex · Head of fibula · Neck · Popliteal line · Fibula · Tibia · Medial malleolus · Lateral malleolus · **Posterior**

Figure 11.5   The tibia and fibula, anterior and posterior views.

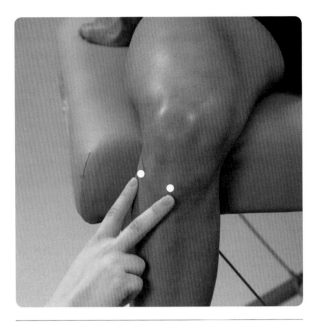

Figure 11.6 Pointing out the tibial tuberosity and head of the fibula.

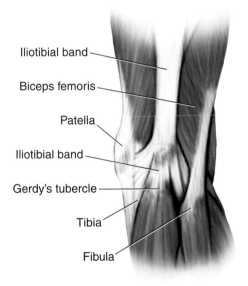

Figure 11.7 Gerdy's tubercle and the iliotibial band.

are where the condyles of the femur articulate with the tibia to form the knee joint (figure 11.8). The large prominence of bone between the two condylar surfaces is known as the **intercondylar eminence,** and the smaller prominences alongside of the intercondylar eminence are known as the **medial** and **lateral intercondylar tubercles** (see figure 11.5).

The final bone of the knee joint is vital to proper movement of the joint. The **patella** (kneecap) is the largest sesamoid (free-floating; see chapter 1)

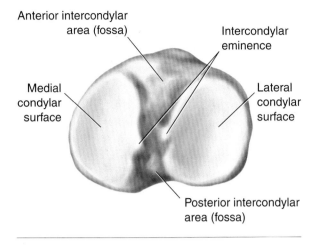

Figure 11.8 The tibia, superior view.

bone in the body (figure 11.9). It is not directly attached to other bones to form a joint. It is embedded in the tendon of insertion of the anterior knee muscle group known as the quadriceps femoris (discussed later in the section on muscles). This muscle group inserts on the patella through

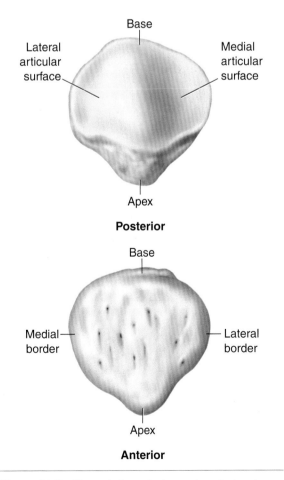

Figure 11.9 The patella, anterior and posterior views.

a broad fibrous sheath, which in turn attaches to the tibial tuberosity through the patellar ligament and tendon. Not only does the patella protect the structures beneath it, but it also changes the "angle-of-pull" of the quadriceps femoris to create a greater rotary force (in flexion and extension) of the knee joint as compared to the stabilizing force of the quadriceps femoris that pulls the tibia into the femur. The proximal end of the patella is known as the **base,** and the distal end is known as the **apex.** The posterior surface has **lateral** and **medial articular surfaces** that are covered with articular cartilage and articulate with the medial and lateral condyles of the femur to form the patellofemoral articulation.

Ligaments of the Knee

Like all synovial joints, the knee joint has a **capsular ligament.** This capsular ligament is unlike others because it consists of portions of other ligaments and fibrous expansions of other structures that cross the knee joint and become part of the capsule (figures 11.10 and 11.11). The components of the capsule include portions of the **medial** and **lateral collateral ligaments,** fibrous expansions of the quadriceps femoris, the iliotibial band, the vastus muscles, the sartorius muscle, and the semimembranosus muscle. Probably the best illustration of the pure capsular ligament exists when one observes the posterior **(popliteal space)** of the knee joint.

The medial collateral ligament (MCL) and lateral collateral ligament (LCL) of the knee provide stability to either side of the joint, essentially preventing abduction and adduction of the joint and making it a uniaxial joint that flexes and extends in the sagittal plane. The medial collateral ligament runs from the medial condyle of the femur to the medial condyle of the tibia, with some deep fibers attaching to the medial meniscus (discussed later in this chapter). The lateral collateral ligament runs from the lateral condyle of the femur to the head of the fibula. Note that the lateral collateral ligament, unlike the medial collateral

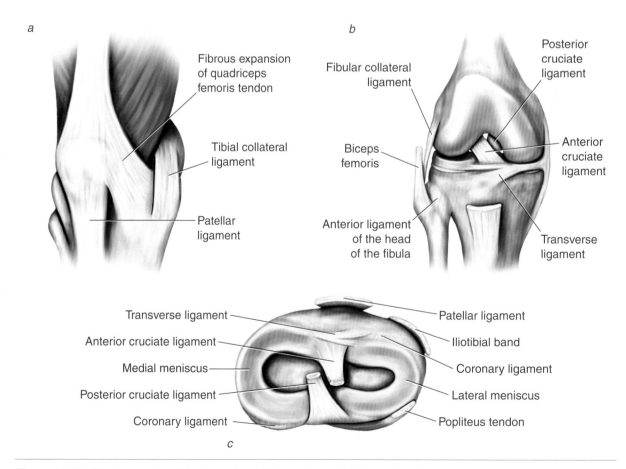

Figure 11.10   The ligaments of the knee, *(a and b)* anterior and *(c)* superior views.

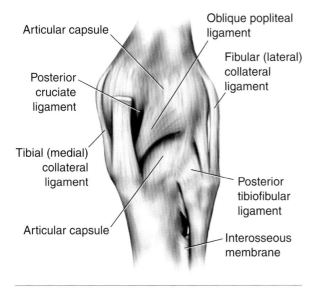

Articular capsule
Oblique popliteal ligament
Posterior cruciate ligament
Fibular (lateral) collateral ligament
Tibial (medial) collateral ligament
Posterior tibiofibular ligament
Articular capsule
Interosseous membrane

Figure 11.11 The ligaments of the knee, posterior view.

ligament, does *not* have fibers attaching to the lateral meniscus.

Hands on . . . As you sit in a chair, place the ankle of one leg on top of the other leg's knee joint. Palpate the lateral condyles of both the femur and tibia of the leg crossed on top of the knee of the other leg. Find the joint space between the condyles. Move your finger anterior and posterior through the joint line until you feel a cordlike structure. This structure is the lateral collateral ligament. Now consider the role of the medial and lateral collateral ligaments in the mechanics of the knee. Remember that the knee normally only flexes and extends. What ligament would come

under stress if a force was applied to the lateral side of the knee joint (often referred to as a *valgus* force), forcing the joint into abduction (i.e., increasing the space between the tibia and the femur on the medial side of the joint)? (See figure 11.12*a* for the answer.) What ligament would come under stress if a force was applied to the medial side of the knee joint (a *varus* force), forcing the joint into adduction (increasing the space between the tibia and the femur on the lateral side of the joint)? (See figure 11.12*b*.) These stresses placed on the MCL and LCL ligaments are commonly the mechanism of sprains of the knee joint.

In the middle of the knee joint are two ligaments known as the **anterior cruciate ligament** (ACL) and the **posterior cruciate ligament** (PCL) (see figures 11.10 and 11.11). The term *cruciate* means "cross," and these two ligaments actually cross each other as they pass through the middle of the knee joint. The anterior cruciate ligament runs from just anterior to the intercondylar eminence of the tibia to the posterior medial surface of the lateral condyle of the femur. The primary function of the anterior cruciate ligament is to prevent anterior displacement of the tibia off the distal end of the femur. This ligament, along with the resistance of the posterior muscles crossing the knee joint, prevents the normal knee from hyperextending. The posterior cruciate ligament runs from just posterior to the intercondylar eminence of the tibia to the anterior portion of the medial surface of the medial condyle of the femur. The primary function of the posterior cruciate ligament is to

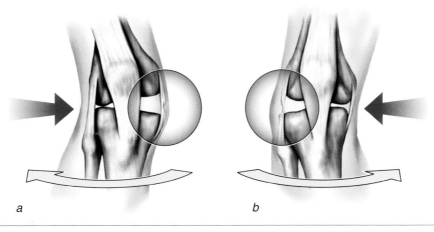

a

b

Figure 11.12 *(a)* A lateral force to the right knee placing a valgus stress on the medial collateral ligament. *(b)* A medial force to the right knee placing a varus stress on the lateral collateral ligament.

Focus on . . . THE ANTERIOR CRUCIATE LIGAMENT

The ACL has received much attention recently in athletics. Many sport activities apply external and internal forces to the knee joint structures including the ACL. There also has been a great amount of interest in the difference in the numbers of ACL injuries between the female and male knee. Just a few of the theories being expressed regarding the larger number of female ACL injuries include: lower extremity malalignment (wider female pelvis, increased Q-angle); ligament laxity; quadriceps/hamstring strength ratio imbalance; smaller female intercondylar (femoral) notch and femoral condyle size; the effects of estrogen on ligament tissue; maturation rates and the effect on landing positions (straight-knee landing, one-step stop landing); use of ankle braces transferring stress to the knee; and the interface between footwear and playing surfaces. These are some of the anatomical theories currently advanced regarding the differences between the female and male knees. Other factors considered are the intensity of training, the type of activity, coaching techniques and many more theories that could fill an entire textbook. Simply said, physicians, biomechanists, physical therapists, athletic trainers, and other interested parties continue to search for the reasons for the differences in the number of ACL injuries between female and male athletes.

prevent posterior displacement of the tibia off the distal end of the femur. Excessive squatting may place the knee joint into such a degree of flexion that this ligament comes under stress.

Three ligaments of the knee joint are found exclusively on the posterior aspect of the joint: the **oblique popliteal ligament,** the **arcuate ligament,** and the **ligament of Wrisberg** (figure 11.13). The oblique popliteal ligament runs from the posterior aspect of the lateral condyle of the

femur to the posterior edge of the medial condyle of the tibia. The arcuate ligament runs from the posterior aspect of the lateral condyle of the femur to the posterior surface of the capsular ligament. The ligament of Wrisberg runs between the posterior horn of the lateral meniscus (see the section on menisci later in this chapter) and the posterior aspect of the medial condyle of the femur.

On the anterior side of the knee, running between the apex of the patella and the tibial

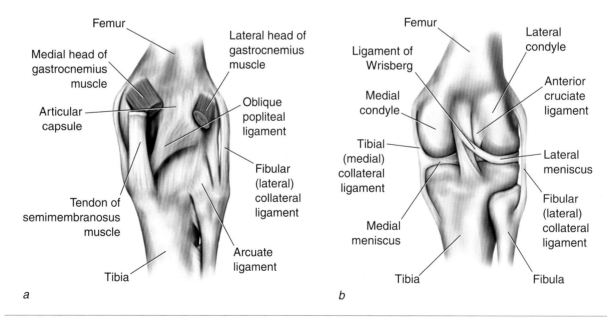

Figure 11.13   *(a)* The oblique popliteal ligament, the arcuate ligament, and *(b)* the ligament of Wrisberg, posterior views.

tuberosity, is the **patellar ligament** (figures 11.10 and 11.14). Because the patella is embedded in the patellar tendon, some anatomists consider this structure to be an extension of the quadriceps femoris tendon of insertion. Others, however, label this structure the patellar ligament because it ties bone to bone (patella to tibia). In this text, the

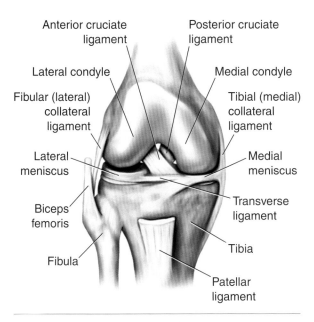

Anterior cruciate ligament
Posterior cruciate ligament
Lateral condyle
Medial condyle
Fibular (lateral) collateral ligament
Tibial (medial) collateral ligament
Lateral meniscus
Medial meniscus
Biceps femoris
Transverse ligament
Fibula
Tibia
Patellar ligament

Figure 11.14 The ligaments of the knee, anterior view.

structure is known as the patellar ligament. Even though some anatomists hold that the quadriceps tendon inserts on the tibial tuberosity, the simple approach is that the quadriceps muscles attach to the patella and the patella attaches to the tibial tuberosity via the patellar ligament.

Two additional ligaments of the knee joint are unique in that they do not tie bone to bone, the normal function of ligaments. The **coronary ligament,** actually a portion of the capsular ligament, is responsible for connecting the outer edges of the menisci (see the next section) to the proximal end of the tibia (see figure 11.10). The **transverse ligament** runs between the anterior horns of the medial and lateral menisci (figure 11.14). This ligament prevents the anterior horn of each meniscus from moving forward when the knee joint moves into extension and the condylar surfaces of both the femur and the tibia exert pressure on the menisci.

Menisci of the Knee

Two semilunar (crescent-shaped), fibrocartilaginous structures sit on the proximal end of the tibia, on the medial and lateral condylar surfaces. These structures are known as the lateral **meniscus,** which is nearly circular in shape, and the semicircular medial

---

Focus on . . . DISRUPTION OF THE MENISCUS

The menisci, more the medial than the lateral, often come under stress through athletic participation. Stress to the deep fibers of the MCL that attach to the medial meniscus could disrupt the meniscus. Excessive rotation of the femur on a fixed tibia (rigidly planted or pinned on the ground or playing surface) could cause stress to the menisci. Two of the most common disruptions of the menisci are called parrot-beak and bucket-handle tears (figure 11.15). The most common, the bucket-handle tear, actually causes the middle portion of the meniscus (between the anterior and posterior horns) to split, causing the inner portion of the meniscus to look like the handle of a bucket as it separates from the main body of the meniscus. Other types of tears of the menisci also occur. Equipment design, playing surfaces, footwear, and specific exercises are all areas of interest for decreasing the incidence of this serious condition.

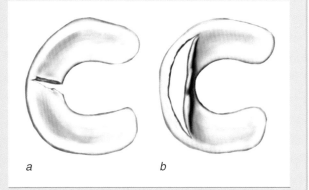

a   *b*

Figure 11.15 Two common tears of the meniscus of the knee: *(a)* parrot-beak tear and *(b)* bucket-handle tear.

meniscus. These structures help deepen the condylar surfaces of the tibia where the condyles of the femur articulate. Their outer borders are thick and convex and are attached to the tibia by the coronary ligament. The inner edges of the menisci are paper thin and lie freely (unattached) on the floor of the condylar surfaces of the tibia. The inner surfaces of the menisci are concave to accommodate the condyles of the femur. The anterior and posterior aspects of each meniscus are often referred to as the **anterior** and **posterior horns** of the meniscus.

Movements of the Knee and Lower Leg

Because the knee joint is a uniaxial joint capable of movement in the sagittal plane about an ever-changing frontal horizontal axis, the only two movements the joint is capable of are flexion and extension. However, because of the sizes and shapes of the femoral condyles and the soft tissue configurations, when the knee flexes and extends, the lower leg (tibia, fibula) rotates. When the knee extends, the leg externally rotates. When the knee flexes, the leg internally rotates. As the knee joint "locks" into extension and "unlocks" when moving into flexion, the rotation of the leg is difficult to see when weight is not being borne by the leg. The locking into full extension is often referred to as the "screw home" movement.

Hands on . . . Sit in a chair with your knees flexed to 90°. With your shoes off, place your feet on a smooth surface (e.g., tile or cement—not carpeting). Make a fist with one hand and place it between your knees. Internally rotate your lower legs so that your toes (first metatarsophalangeal joints; see chapter 12) on both feet push against each other. As you internally rotate your lower legs to push your feet against each other, with your free hand feel the semimembranosus (medial thigh, just proximal to the knee) muscles posterior to your knee joints. These muscles (and those of the pes anserinus—see later in this chapter) are contracting to cause the internal rotation of the lower leg. Now, from the same starting position, have your partner place his or her feet along the sides or the lateral aspect of both of your feet. Find the biceps femoris tendon on the posterior lateral aspect of your knee

with your free hand. Now try to externally rotate your lower legs to force your feet to push your partner's feet away. What did you feel? As the muscles that cross the knee joint are reviewed, their bony attachments will help identify what, if any, movement of the leg they perform.

Muscles of the Knee and Lower Leg

The muscles crossing the knee joint can easily be divided into those crossing the joint anteriorly and those posteriorly. The anterior group is reviewed first.

Anterior Muscles

Because of their position anterior to the knee joint (figure 11.16), the logical conclusion is that these muscles extend the joint. This is primarily true, but other functions are also involved, as noted in the following discussions.

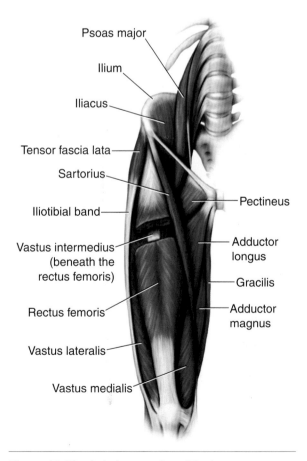

Figure 11.16   Anterior muscles of the knee.

> **Sartorius:** This muscle, the longest in the body, was examined in the hip chapter because it is a flexor of the hip. The sartorius originates on the anterior superior iliac spine (ASIS), crosses the hip joint, passes posterior to the medial condyle of the femur, and inserts just inferior to the proximal end of the medial surface of the tibia (figure 11.16). At the knee joint, the sartorius flexes the knee and internally rotates the lower leg.

The **quadriceps femoris,** frequently referred to as the "quads," is a group of four muscles (figures 11.16 and 11.17). One of these muscles, the rectus femoris, crosses both the knee and the hip joint and was reviewed in the hip chapter as it pertained to hip joint action. Here we look at its function as a knee extensor. The other three muscles–the vastus intermedius, vastus medius, and vastus lateralis–cross only the knee joint and have only one function: extension at the knee.

> **Rectus femoris:** This is the one quad muscle that crosses both the hip and knee joints. It is the most superficial of the anterior thigh muscles. Its straight head originates on the anterior inferior iliac spine, its reflected head originates on the acetabulum, and it inserts on the base of the patella (figure 11.16). The rectus femoris is an extensor of the knee joint.

> **Vastus lateralis:** The largest of the three vastus muscles, the vastus lateralis originates on the proximal half of the linea aspera, the **intertrochanteric line,** and the greater trochanter of the femur and inserts on the lateral border of the patella. The vastus lateralis extends the knee joint.

> **Vastus medialis:** Originating on the medial lip of the linea aspera, the vastus medialis inserts on the medial border of the patella. The vastus medialis extends the knee joint.

Hands on . . . With resistance to the lower leg (from either a partner, an iron boot, or a flexion–extension machine), extend your knee (see figure 11.18). Observe and palpate the medial and lateral aspects of your thigh. Identify the musculature you are observing.

Figure 11.17 Viewing the quadriceps femoris muscle group.

Figure 11.18 Locating the rectus femoris.

Focus On . . . **Q-ANGLE AND PFS**

Chapter 7 pointed out the differences in the shape of the female and male pelvis. If you draw a line through the patella and the tibial tuberosity and another line between the patella and the anterior superior iliac spine (ASIS), you will note an angle between these two lines (figure 11.19). This angle is referred to as the *Q-angle*. The Q-angle is defined as the angle between the line of the quadriceps muscle pull and the line of insertion of the patellar tendon. In females this angle normally is between 15° and 20°, whereas in males it ranges from 10° to 15°. A Q-angle greater than 20° can result in any of a multitude of problems causing pain in and around the patellofemoral joint. This all-inclusive evaluation of "pain" is often identified as patellofemoral syndrome (PFS). As mentioned, there are multiple reasons that may cause anterior knee joint pain, with an excessive Q-angle as a possible contributing factor. We will not go into an extended medical diagnosis in an entry-level anatomy textbook, but suffice it to say that those experiencing PFS often complain about pain while climbing stairs. Consider the relationship of the patella and the femur when the knee is moved during stair climbing.

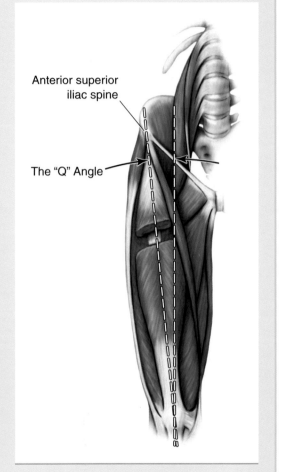

Figure 11.19   The Q-angle.

> **Vastus intermedius:** Beneath the rectus femoris lies the vastus intermedius. This muscle originates on the proximal two thirds of the anterior surface of the femur and inserts on the inferior surface of the patella. The vastus intermedius extends the knee joint.

Beneath the quadriceps femoris muscles lies another anterior muscle, the genu articularis.

> **Genu articularis:** This muscle, deep beneath the vastus intermedius, originates on the anterior surface of the femur just proximal to the condyles and inserts, not on another bone, but on the **synovial membrane** of the knee joint. As the knee moves into extension, this muscle contracts, pulling the articular capsule of the knee proximally to prevent the synovial membrane from becoming impinged between the femur, the patella, and the tibia (figure 11.20).

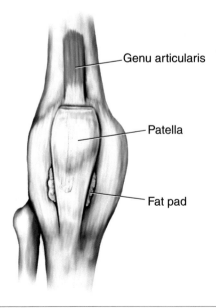

Figure 11.20   The genu articularis.

Posterior Muscles

The posterior muscles of the knee joint (figure 11.21) include muscles, such as the hamstrings and the gracilis, whose actions at the hip joint are reviewed in the discussion of muscles in chapter 10. Here we look at their actions at the knee joint and also at the other muscles that cross the knee posteriorly.

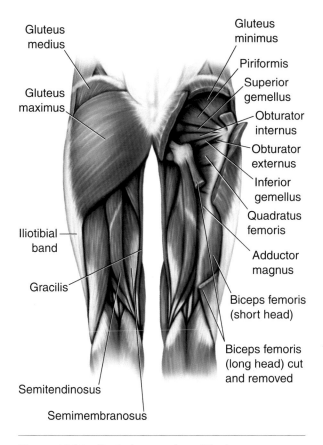

Figure 11.21 Posterior muscles of the knee.

> **Biceps femoris:** This hamstring muscle has two heads: One, the long head, originates on the ischial tuberosity, and the other, the short head, originates on the lateral aspect of the linea aspera (figure 11.21). The muscle inserts on the head of the fibula. The biceps femoris flexes the knee, and as the knee reaches active full flexion, it externally rotates the lower leg.

> **Semitendinosus:** The second hamstring muscle, the semitendinosus, originates on the ischial tuberosity and inserts on the proximal aspect of the medial tibia (figure 11.21). The semitendinosus flexes the knee and internally rotates the lower leg.

> **Semimembranosus:** The third hamstring muscle, the semimembranosus, originates on the ischial tuberosity and inserts on the posterior medial aspect of the medial condyle of the tibia (figure 11.21). The semimembranosus flexes the knee and assists with internal rotation of the lower leg.

> **Gracilis:** The gracilis is the only adductor muscle of the hip joint that also crosses the knee joint. It originates on the inferior surface of the pubic symphysis, runs posterior to the medial condyle of the femur, and inserts just posterior to the medial aspect of the proximal end of the tibia (figure 11.21). The gracilis flexes the knee joint and internally rotates the lower leg.

The gracilis, semitendinosus, and sartorius all insert in the same general area, just below the proximal end of the tibia on its medial aspect. The insertion of the three closely grouped tendons of insertion is commonly identified as the **pes anserinus** (figure 11.22). All three components of the pes anserinus flex the knee joint and internally rotate the lower leg.

> **Popliteus:** Diagonally crossing the popliteal space of the knee joint, the popliteus runs between the lateral aspect of the lateral condyle of the femur and the popliteal line on the proximal third of the posterior surface of the tibia (figure 11.23).

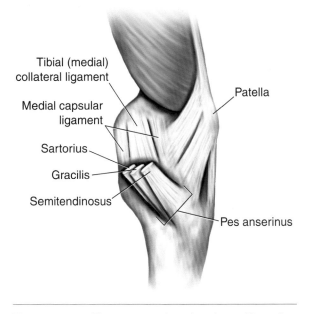

Figure 11.22 The pes anserinus (tendons of insertion of the sartorius, gracilis, and semitendinosus).

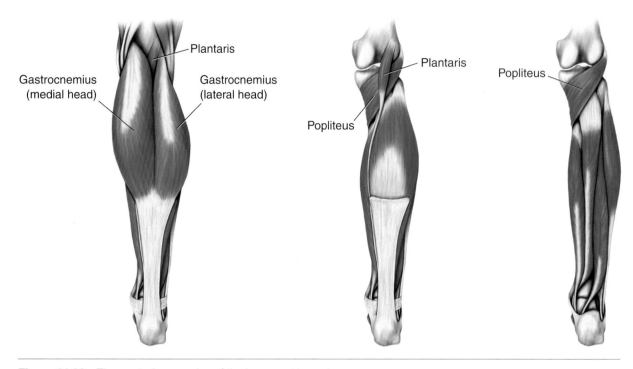

Figure 11.23   The posterior muscles of the knee and lower leg.

Observe the arrangement of the fibers of this muscle. Besides flexion of the knee, what obvious motion of the lower leg is likely to occur when this muscle contracts? (Answer: internal rotation)

> **Iliotibial band:** This structure (figure 11.24*a*), also discussed in the hip chapter, is a combination of the **gluteus maximus** and **tensor fascia lata** tendons of insertion (figure 11.21). It crosses the knee in the area of the lateral condyle of the femur and inserts onto a bony prominence just inferior and anterior to the lateral condyle of the tibia known as Gerdy's tubercle (see figure 11.7). This structure both flexes and extends the knee joint, depending on the angle of the knee joint at any particular moment. When the knee joint is between full extension and 10° to 15° of flexion, the iliotibial band is anterior to the lateral femoral condyle and assists with extension of the knee joint (figure 11.24*b*). As the knee joint continues to flex beyond 10° to 15°, the iliotibial band shifts to a position posterior to the lateral femoral condyle and becomes a flexor of the knee joint (figure 11.24*c*).

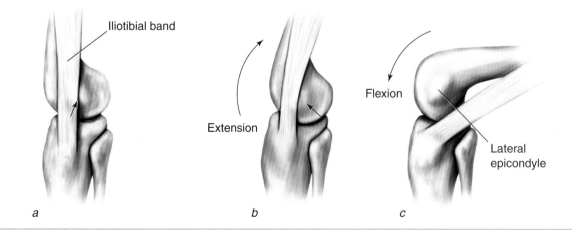

Figure 11.24   *(a)* The iliotibial band. *(b)* The iliotibial band as the knee joint approaches full extension. *(c)* The iliotibial band as the knee joint moves to flexion.

Hands on . . . Palpate the lateral side of your knee joint. Move your hand proximally, just above the lateral condyle of the femur. Flex and extend your knee joint and feel the iliotibial band move anterior and posterior to the femoral condyle.

Two muscles of the lower leg originate above the knee joint and play a role in knee joint function, although their primary functions involve the ankle joint.

> **Gastrocnemius:** This muscle has two heads: one originating on the posterior aspect of the lateral condyle of the femur and the other originating on the posterior aspect of the medial condyle of the femur (figure 11.23). Both heads combine into a single tendon of insertion that attaches to the **calcaneus.** The gastrocnemius flexes the knee joint. In the rare instance where the gastrocnemius cannot perform its primary function (plantar flexion of the ankle joint) because the foot is held in a fixed position and cannot move, contraction of the gastrocnemius can cause extension of the knee joint.

> **Plantaris:** This short-bellied muscle with a long tendon of insertion originates on the lateral linea aspera and the oblique popliteal ligament and inserts on the calcaneus (figures 11.23). The plantaris muscle assists with knee flexion but is of little importance in humans compared with the other muscles that perform the same function.

REVIEW OF TERMINOLOGY

The following terms were discussed in this chapter. Define or describe each term, and where appropriate, identify the location of the named structure either on your body or in an appropriate illustration.

adductor tubercle	lateral collateral ligament	patellofemoral joint
anterior cruciate ligament	lateral condylar surface (of the tibia)	pes anserinus
anterior horn	lateral condyle of the femur	plantaris
apex of the fibula	lateral condyle of the tibia	popliteal groove
apex of the patella	lateral epicondylar ridge (of the femur)	popliteal line
arcuate ligament	lateral epicondyle (of the femur)	popliteal space
base of the patella	lateral intercondylar tubercle	popliteal surface of the femur
biceps femoris	ligament of Wrisberg	popliteal surface of the tibia
calcaneus	medial articular surface (of the patella)	popliteus
capsular ligament	medial collateral ligament	posterior cruciate ligament
coronary ligament	medial condylar surface (of the tibia)	posterior horn
femoral notch	medial condyle of the femur	quadriceps femoris
femur	medial condyle of the tibia	rectus femoris
fibula	medial epicondylar ridge (of the femur)	sartorius
gastrocnemius		semimembranosus
genu articularis	medial epicondyle (of the femur)	semitendinosus
Gerdy's tubercle	medial intercondylar tubercle	synovial membrane
gluteus maximus	meniscus	tensor fascia lata
gracilis	neck of the fibula	tibia
head of the fibula	oblique popliteal ligament	tibial tuberosity
iliotibial band	patella	transverse ligament
intercondylar eminence	patellar facet	vastus intermedius
intercondylar notch	patellar ligament	vastus lateralis
intertrochanteric line		vastus medialis
lateral articular surface (of the patella)		

SUGGESTED LEARNING ACTIVITIES

1. Lying supine on a table with your lower legs off the end of the table, extend both knee joints to full extension.
   a. What muscles performed this action?
   b. Which of these muscles played the most prominent role in the joint action?
   c. Did you have difficulty fully extending your knee joints? If so, why?

2. Sitting up on a table with your lower legs off the end of the table, extend both knee joints to full extension.
   a. What muscles performed this action?
   b. Which of these muscles played the most prominent role in the joint action?
   c. Did you have difficulty fully extending your knee joints? If so, why?

3. Sitting up on a table with your lower legs off the end of the table and your feet next to each other, extend both knee joints to full extension. As your knees moved to full extension, what did the lower legs (and the relationship of your feet to each other) do?

4. Lying prone with your entire body on a table, bring your heels to your buttocks.
   a. What muscles performed this action of the knee joints?
   b. Did your pelvis rise off the table as you performed this activity? If so, why?

MULTIPLE-CHOICE QUESTIONS

1. Which of the following ligaments has deep fibers attaching to a meniscus?
   a. medial collateral
   b. lateral collateral
   c. anterior cruciate
   d. posterior cruciate

2. Which of the following structures attaches to Gerdy's tubercle on the proximal anterior lateral aspect of the tibia?
   a. quadriceps tendon
   b. lateral collateral ligament
   c. lateral meniscus
   d. iliotibial band ligament

3. The popliteal space is found on what aspect of the knee joint?
   a. anterior
   b. posterior
   c. lateral
   d. medial

4. Which of the following ligaments is responsible for preventing posterior displacement of the proximal tibia off the distal end of the femur?
   a. anterior cruciate
   b. posterior cruciate
   c. medial collateral
   d. lateral collateral

5. When the knee joint moves into extension, which muscle becomes the external rotator of the lower leg?
   a. plantaris
   b. popliteus
   c. biceps femoris
   d. rectus femoris

6. Which of the following hamstring muscles is part of the structure known as the pes anserinus?
   a. semitendinosus
   b. semimembranosus
   c. rectus femoris
   d. biceps femoris

7. Which of the following muscles crosses only the knee joint?
   a. sartorius
   b. biceps femoris
   c. rectus femoris
   d. popliteus

8. Which of the following muscles, although considered a weak flexor of the knee, is really of little significance in human anatomy?
   a. popliteus
   b. plantaris
   c. biceps femoris
   d. sartorius

9. Which of the following hip adductor muscles also crosses the knee joint?
    a. pectineus
    b. adductor magnus
    c. gracilis
    d. adductor longus

10. Which of the following muscles is not considered a primary flexor of the knee?
    a. semitendinosus
    b. semimembranosus
    c. biceps femoris
    d. rectus femoris

11. Which of the following muscles is the longest muscle in the human body?
    a. sartorius
    b. rectus femoris
    c. semitendinosus
    d. tensor fascia lata

12. Which of the following muscle groups contains the largest sesamoid bone in the body within its tendon of insertion?
    a. iliopsoas
    b. quadriceps
    c. hamstrings
    d. adductors

13. Which of the following muscles is considered the lateral hamstring muscle?
    a. semitendinosus
    b. semimembranosus
    c. biceps femoris
    d. rectus femoris

14. Which of the following muscles is found beneath the rectus femoris muscle?
    a. vastus lateralis
    b. vastus intermedius
    c. vastus medialis
    d. vastus femoris

FILL-IN-THE-BLANK QUESTIONS

1. The most distal aspect of the femur bone is its _____.

2. The medial and lateral menisci are attached to the _____.

3. The ligament attaching the anterior horns of the medial and lateral meniscus to prevent meniscal distortion during knee extension is the _____ _____ ligament.

4. The gastrocnemius muscle originates on the _____.

5. The ligament designed to prevent forward displacement of the tibia off the distal end of the femur is the _____ ligament.

6. A joint defined as a modified hinge joint of the lower extremity is the _____ joint.

7. The iliotibial band consists of the combined tendons from the tensor fascia lata and the _____ _____.

CHAPTER TWELVE

The Lower Leg, Ankle, and Foot

This chapter may seem familiar because the lower leg, ankle, and foot are very similar anatomically to the forearm, wrist, and hand. The bones, ligaments, and muscles may or may not have similar names, but their structures are similar. In contrast to the upper extremity, the lower extremity is constructed to bear the weight of the body and absorb the force applied by that body weight with each foot strike.

Bones of the Lower Leg

The two bones of the lower leg are the **tibia** (medially) and the **fibula** (laterally). Much debate has been conducted over the weight-bearing aspects of both bones, but in this text, we assume that the tibia is the major weight-bearing bone and that the fibula has little or no weight-bearing function.

The prominent bony markings of the proximal end of these bones were presented in chapter 11 on the knee. At the distal ends of the shafts of both the tibia and the fibula are large bumps known as **malleoli** (figures 12.1and 12.2). The prominence at the distal end of the fibula is known as the **lateral malleolus,** and the prominence at the distal end of the tibia is known as the **medial malleolus.** Palpate these bumps on either side of the distal end of your lower leg and note which one is more distal than the other. This is important to remember later, when the fundamental movements of the ankle joint are discussed.

On the lateral surface of the tibia, just proximal to the distal end, is a notch known as the **fibular notch of the tibia,** where the tibia and fibula articulate to form the distal **tibiofibular joint.**

The distal surfaces of both the tibia and the fibula have facets (or smooth surfaces) that articulate with the talus, one of the bones of the foot (figure 12.3).

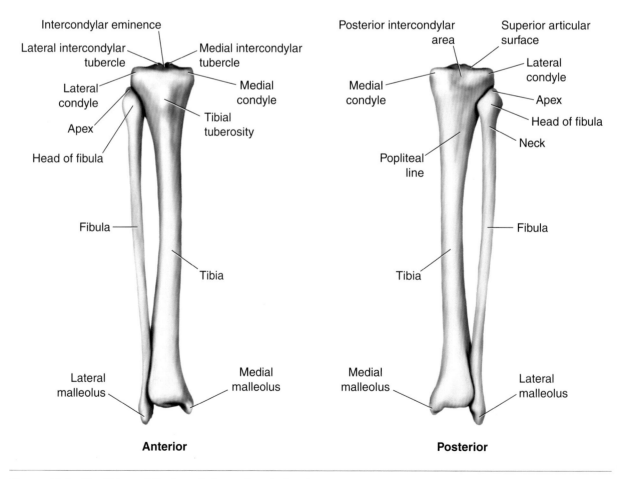

Figure 12.1　The tibia and fibula, anterior and posterior views.

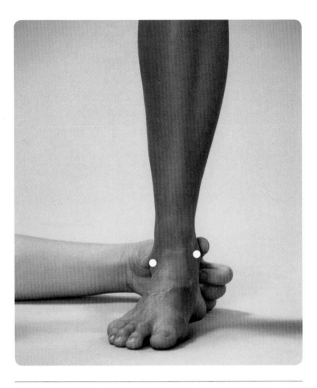

Figure 12.2 Locating the malleoli.

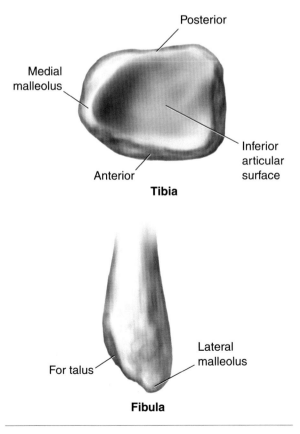

Figure 12.3 The distal ends of the fibula (posterior view) and the tibia (inferior view).

Bones of the Foot

The 26 bones of the foot (figure 12.4) are usually separated into three distinct segments: the **forefoot** (19), the **midfoot** (5), and the **hindfoot** (2).

The forefoot consists of 14 **phalanges:** three per toe (proximal, middle, and distal phalanges), except for the great toe, which has only a proximal and a distal phalanx. The phalanges and the metatarsal bone of the first, or great, toe are larger than those of the other four toes (second, middle, fourth, and little) for a specific purpose. When the foot bears the weight of the body, as in walking, the great toe must bear most of the weight. This is the same arrangement we found in the hand: three phalanges per finger and only two in the thumb. The other five bones of the forefoot are the five long bones of the foot, known as the **metatarsal bones.** Again, these bones are similar to the five metacarpal bones of the hand. The metatarsal bones consist of a **head** (distal end), a **shaft,** and a **base** (proximal end). The most prominent of these structures is the base of the fifth metatarsal bone.

Hands on . . . Run your finger along the lateral edge of your foot and palpate a rather large bump about half to two thirds of the length of your foot from your fifth (little) toe. This is the base (also known as the tuberosity) of the fifth metatarsal bone (see figure 12.4).

The remaining seven bones of the foot are collectively known as the **tarsal bones.** Five of these tarsal bones (cuboid, navicular, and the medial, intermediate, and lateral cuneiforms) make up the midfoot, and the remaining two tarsal bones (talus and calcaneus) make up the hindfoot.

Just proximal to the medial metatarsal bones are three tarsal bones known as the **cuneiform bones** (figures 12.4 and 12.5). These are often referred to as the first, second, and third cuneiforms, but it is easier to remember them by their anatomical position: the medial, intermediate, and lateral cuneiform bones. Proximal to the cuneiform bones is the fourth midfoot bone, the **navicular bone.** As in the wrist, the navicular bone is also known as the **scaphoid.** The fifth tarsal bone of the midfoot is also the most lateral of the five midfoot tarsals: the **cuboid bone.** Careful examination of this bone reveals that its lateral border is concave. This groove provides a space for a tendon of the

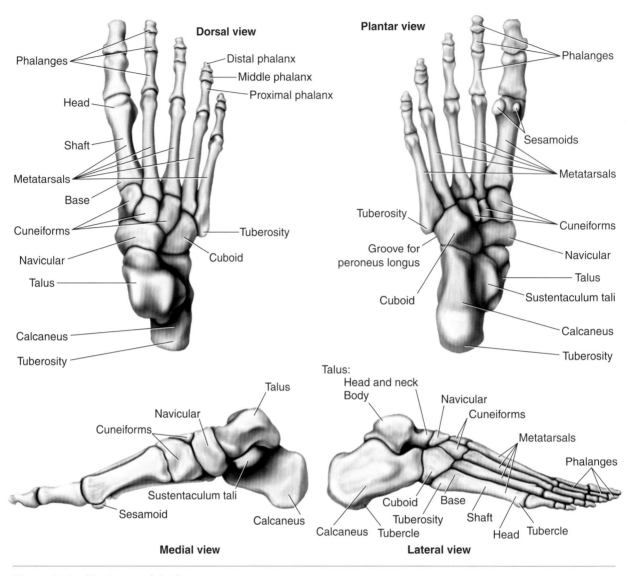

Figure 12.4 The bones of the foot.

foot and ankle (the peroneus longus, discussed later in this chapter).

The final two tarsal bones, the **talus** and the **calcaneus,** make up the bones of the hindfoot (figures 12.4, 12.6 and 12.7). These two bones articulate with the tibia and fibula, similarly to how the scaphoid and lunate bones articulate with the ulna and radius of the forearm. The talus literally sits on top of (proximal to) the calcaneus, which is often referred to as the heel bone and is the largest of the tarsal bones. Note on the posterior view of the calcaneus (figure 12.8) a bony prominence extending medially from the superior surface of the calcaneus. This prominence, the **sustentaculum tali,** serves as a platform for a portion of the talus to sit on. These two bones, articulating together and with the lower-leg bones, form the two joints (talocrural and talocalcaneal) that we refer to as the **ankle joint.** Movement in these two joints (two movements in each joint) result in the four movements attributable to the ankle joint. These joints and movements are further discussed later in this chapter.

On the plantar surface of the head of the first metatarsal bone are two **sesamoid** bones (see figure 12.4), which are not typically counted with the other 26 bones of the foot. Like the patella and sesamoid bones in general, by definition, these sesamoid bones are embedded in the tendon of a muscle and are free-floating. They not only protect the structures superior to them but also provide a biomechanical advantage for the function of the muscle in which they are embedded.

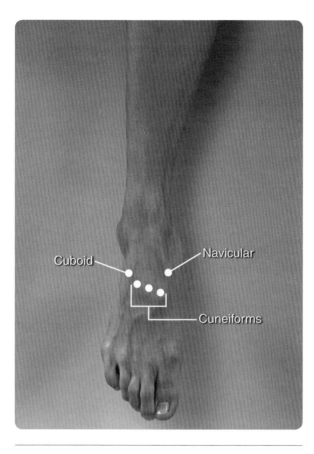

Figure 12.5 A dorsal view of the navicular, cuneiform, and cuboid bones.

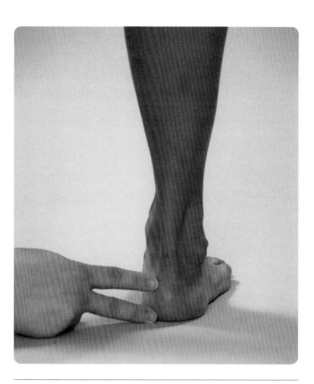

Figure 12.7 Identifying the calcaneus.

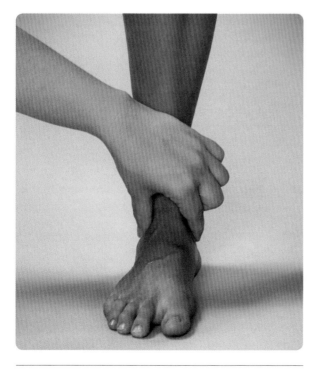

Figure 12.6 Locating the talus.

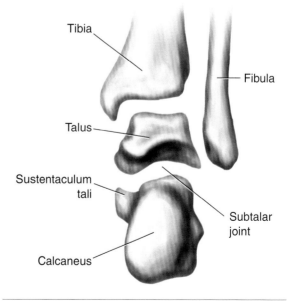

Figure 12.8 Posterior view of the ankle, illustrating the sustentaculum tali of the calcaneus and the subtalar joint.

Ligaments and Joints of the Ankle and Foot

In the following sections, we examine first the joints and ligaments of the ankle and then those of the foot. Finally, we look at the arches of the feet, essential to any discussion of the foot.

The Ankle

The ankle is not really a single well-defined joint like many other articulations throughout the human body. Some authors refer to the *ankle joint complex* because there is more than one joint where the movement that we commonly refer to as ankle joint motion takes place (see page 218).

The major ligaments at the distal end of the leg are the **interosseous ligament (interosseous membrane)** (figures 12.9 and 11.11; similar to the interosseous ligament of the forearm) found between the medial border of the shaft of the fibula and the lateral border of the shaft of the tibia (from proximal to distal ends) and the **anterior** and **posterior tibiofibular ligaments** at the distal end of the leg (figure 12.10). The interosseous ligament serves as a source of attachment for numerous anterior and posterior muscles of the lower leg.

The talocrural aspect of the talocrural–talocalcaneal (ankle) joint is a hinge joint that permits movement in the sagittal plane (dorsiflexion and plantar flexion). In anatomy, the talocrural joint is often referred to as a loosely formed mortise-and-tenon joint. The mortise (recess or hole) of the joint is formed by the lateral malleolus of the fibula and the medial malleolus of the tibia. The tenon (peg) of the joint is the talus, which fits into the mortise (figure 12.11).

In addition to the capsular ligament present in all synovial joints, there are four major ligaments of the ankle joint (figure 12.10). For the most part, the names of ligaments indicate the bones that the ligaments bring together to form articulations.

Medially, the major ankle ligament is known as the **deltoid ligament,** made up of three superficial ligaments and one deep ligament. Superficially, the anterior portion of the deltoid ligament is the **tibionavicular ligament,** the middle portion is the **calcaneotibial ligament,** and the posterior portion is the **posterior talotibial ligament.** The deep ligament of the deltoid ligament is the **anterior talotibial.**

Laterally, the ankle has three major ligamentous structures. The shortest of these three ligaments is the **anterior talofibular** (ATF) **ligament.** It runs from the anterior aspect of the distal end of the lateral malleolus to the talus. The ATF ligament is the most commonly sprained ligament of the ankle joint.

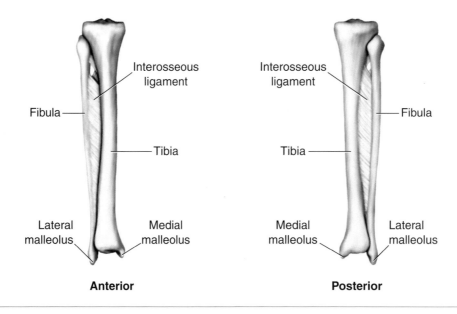

Anterior   **Posterior**

Figure 12.9   The interosseous ligament of the tibiofibular joint.

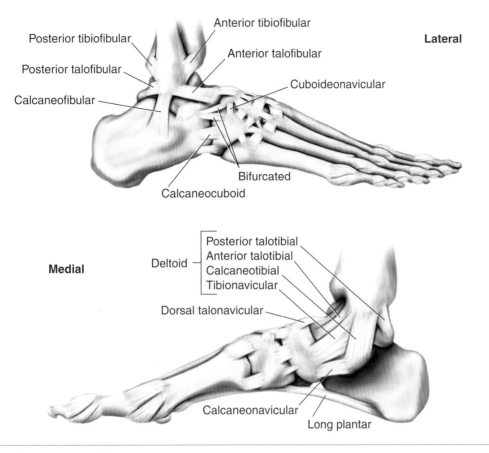

Figure 12.10   The ligaments of the foot and ankle.

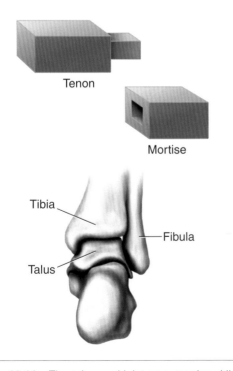

Figure 12.11   The talocrural joint as a mortise- (tibia and fibula) and-tenon (talus) joint.

Hands on . . . Placing your finger just anterior and distal to the end of your fibula, you find a soft depression, the sinus tarsi (*sinus* = "cavity," *tarsi* = "tarsal bone"). The ATF ligament crosses through this area.

The strongest of the three lateral ankle ligaments is the **posterior talofibular ligament,** which runs from the posterior aspect of the lateral malleolus to the talus. The longest of the three lateral ankle joint ligaments is the **calcaneofibular ligament.** It runs from the lateral malleolus of the fibula to the lateral aspect of the calcaneus.

Hands on . . . After observing the anatomical arrangements of the bones and ligaments of the ankle, attempt to turn your ankle inward, toward your other ankle (inversion), and then attempt to turn your ankle outward, away from your other ankle (eversion). Which movement created the greatest motion? What structures limited the movement? Observing these actions and the

Focus on . . . ANKLE JOINT PROBLEMS

Any of the ligaments of the ankle can be sprained, depending on the exact mechanics of the stress. However, the anterior talofibular (ATF) ligament is by far the most commonly sprained ligament of the ankle joint. "Rolling over" an ankle (turning the foot inward excessively toward the other foot) places stress on all lateral ligaments of the ankle, but the ATF in particular is under the greatest amount of stress in this typical ankle sprain. The term *inversion sprain* is frequently used to describe this sprain.

Although much more rare than the common *inversion sprain,* a *high ankle sprain* is caused by the same mechanism. A more anatomically correct term would be a *syndesmosis sprain.* The term *syndesmosis* is another way to define a joint: "an articulation between bones tied together by ligaments." In the case of the ankle joint, when the ankle is excessively inverted, the talus may force against the fibula causing the joint between the fibula and tibia (the tibiofibular syndesmosis) to spread apart. This spreading can possibly sprain the ATF ligament, the posterior tibiofibular ligament, the interosseous membrane, or any combination of these. The spraining of any of these three structures can be defined as a *high ankle sprain.*

Preventing ankle joint problems is important for many people involved in physical activity. Prescribing specific exercises for strengthening or rehabilitating the ankle, designing footwear, using preventive measures (such as taping, wrapping, or bracing), and understanding the effects of specific playing surfaces all rely on one's knowledge of the anatomy of the ankle joint. Prevention strategies rely on knowledge of the types of structures that are stressed during injury, the types of forces involved, how to counteract those potentially damaging forces, and the musculature that can be strengthened to better resist the damaging forces. The disciplines of biomechanics, athletic training, sports medicine, physical therapy, and exercise science can provide the necessary background for addressing this important aspect of sport.

structural arrangement of the bones and ligaments of the ankle should reveal why the ATF is the most likely ligament of the ankle joint to sprain. The terms *inversion* and *eversion* are discussed later in this chapter.

Ligaments of the Foot

The ligaments of the foot (see figures 12.10 and 12.12) can be divided into five groups: the **intertarsal ligaments,** the **tarsometatarsal ligaments,** the **intermetatarsal ligaments,** the **metatarsophalangeal ligaments,** and the **interphalangeal ligaments.**

The intertarsal ligaments tie together the articulations between the tarsal bones of the hindfoot and the bones of the midfoot. The hindfoot joint between the talus and the calcaneus (**talocalcaneal** or **subtalar joint,** figure 12.8) produces medial and lateral gliding movements in the hindfoot that are identified as inversion (movement of the foot toward the midline of the body) and eversion (movement of the foot away from the midline).

Intertarsal ligaments, like the intercarpals of the wrist, join the seven tarsal bones of the foot together. Most of these ligaments are identifiable by their names, which represent the bones they tie together. The talocalcaneonavicular joint has a capsular ligament and a **dorsal talonavicular ligament.** The calcaneocuboid joint has a capsular ligament and the **calcaneocuboid ligament.** The **bifurcated ligament** runs from the calcaneus and divides into fibers that run to the cuboid and to the navicular bones. The **long plantar ligament** runs from the calcaneus to the cuboid, with fibers extending to the bases of the third, fourth, and fifth metatarsal bones (figures 12.10 and 12.12). The long plantar ligament and the **calcaneonavicular ligament** are involved in one of the arches of the foot (figure 12.10). Additional ligaments of the intertarsal group include three dorsal and three plantar **cuneonaviculars;** the dorsal, plantar, and interosseous **cuboideonaviculars;** the dorsal, plantar, and interosseous **intercuneiforms;** and the dorsal, plantar, and interosseous **cuneocuboids.**

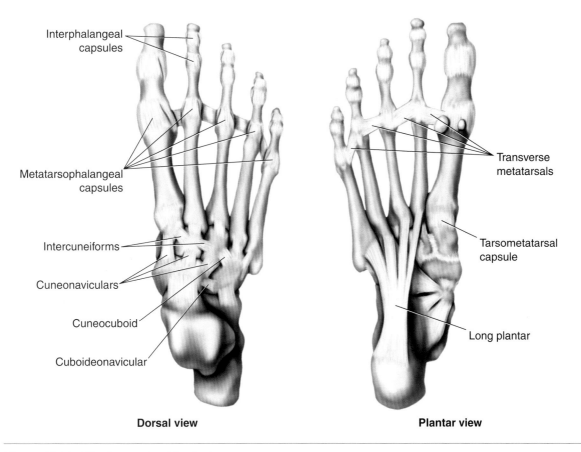

Dorsal view   **Plantar view**

Figure 12.12   The ligaments of the foot.

The tarsometatarsal ligaments are the dorsal and plantar capsular ligaments and **interosseous (collateral) ligaments** (under the capsules on the medial and lateral aspects of the joints) joining the five metatarsal bones to the tarsal bones of the midfoot to form the tarsometatarsal joints (figures 12.10, 12.12 and 12.13). The intermetatarsal ligaments join the bases of the five metatarsal bones together at the tarsometatarsal joints.

The metatarsophalangeal (MP) joints that join the metatarsal bones with the proximal phalanges also have dorsal and plantar capsular ligaments and interosseous (collateral) ligaments under the capsules on the medial and lateral aspects of the joints (figures 12.12 and 12.13). Additionally, there is a **transverse metatarsal ligament** connecting all five heads of the metatarsal bones (figure 12.12). Note that the first and fifth metatarsophalangeal joints are often referred to as the "large ball" and the "small ball" of the foot.

Hands on . . . Observe the plantar surface of your foot and determine for yourself which ball is which.

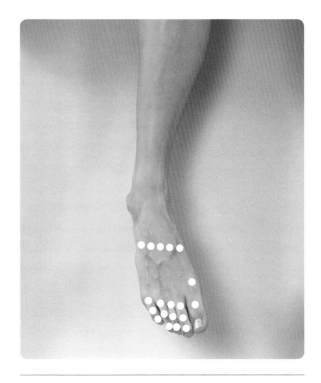

Figure 12.13   Locating the tarsometatarsal, metatarsophalangeal, and interphalangeal joints.

The interphalangeal joints include the four proximal interphalangeal (PIP) joints and four distal interphalangeal (DIP) joints of the four toes and the single interphalangeal (IP) joint of the great toe. The interphalangeal joints are joined by dorsal and plantar capsular ligaments and interosseous (collateral) ligaments.

Arches

Anatomically, an arch is defined as the structures forming a curved or bow-shaped object. The foot contains either two or three arches, depending on one's viewpoint.

The **longitudinal arch** runs from the calcaneus to the heads of the metatarsal bones on the plantar surface of the foot (figure 12.14). The arch is formed by a combination of the shapes of the bones and the ligamentous structures supporting the bones. The space created beneath the arch allows the muscles, tendons, blood vessels, and nerves of the plantar surface of the foot to pass without being crushed against the ground.

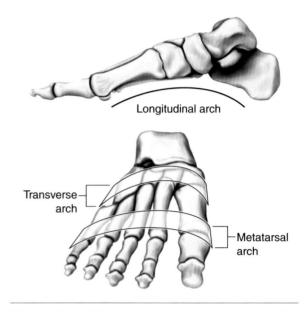

Figure 12.14   The transverse, metatarsal, and longitudinal arches.

The metatarsal bones form an arch (or arches) as the result of their ligamentous attachments to both the phalanges and the tarsal bones. An

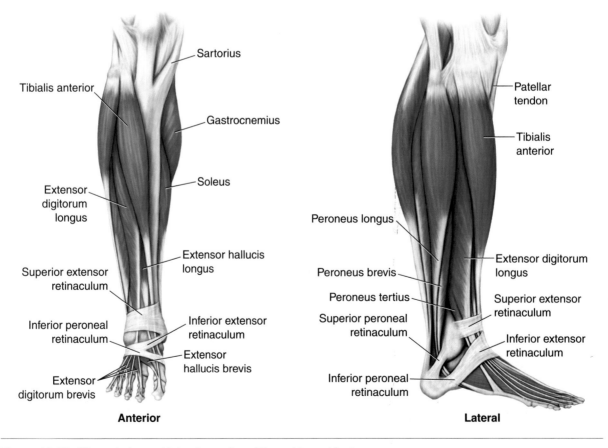

Figure 12.15   The anterior and lateral muscles of the ankle and foot.

anterior view of the foot reveals that the metatarsophalangeal joints form an arch referred to as the **metatarsal arch.** This same arch appears at the other (proximal) end of the metatarsal bones as they articulate with the tarsal bones. Some authors refer to this arch also as the metatarsal arch, whereas others refer to it as the **transverse arch** of the foot (figure 12.14).

Five other structures (similar to the two found in the wrist) in the ankle and foot are the **superior** and **inferior extensor retinacula,** the **superior** and **inferior peroneal retinacula,** and the **flexor retinaculum** (figures 12.15 and 12.16). All these

structures function primarily to keep the tendons of muscles in their appropriate positions.

Movements of the Lower Leg, Ankle, and Foot

Upward movement of the foot toward the anterior leg is known as **dorsiflexion** of the ankle joint. Downward movement of the foot is known as **plantar flexion** (figure 12.17). Lateral (outward) movement of the foot at the talocalcaneal (subtalar) joint produces a movement known as

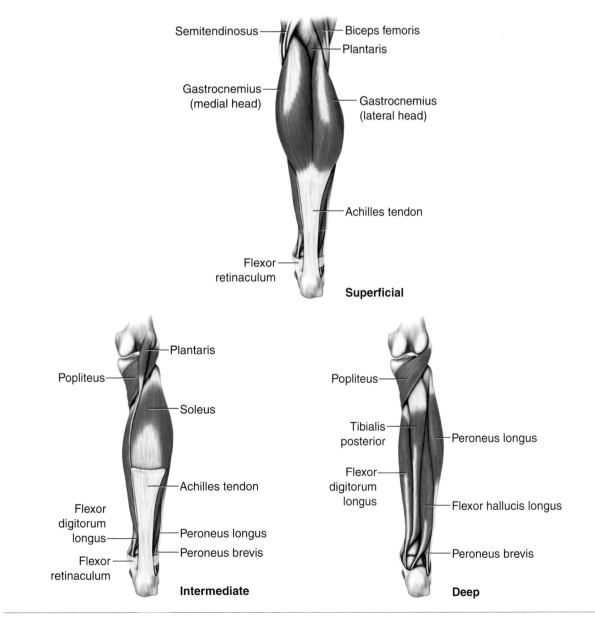

Figure 12.16 The posterior muscles of the ankle and foot.

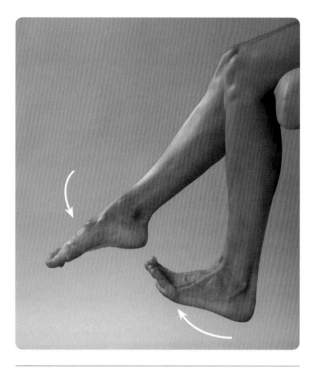

Figure 12.17   Plantar flexion of the right ankle and dorsiflexion of the left.

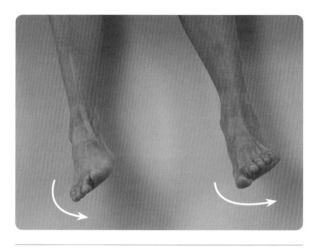

Figure 12.18   Inversion of the right ankle and eversion of the left.

eversion. Medial (inward) movement of the foot at the talocalcaneal joint produces a movement known as **inversion.** The motion between the hindfoot (talus) and midfoot (navicular) bones also contributes to these movements (figure 12.18).

Although the talocrural joint is considered the true ankle joint, most authors cite plantar flexion, dorsiflexion, inversion, and eversion as the four fundamental movements of the ankle joint.

Because plantar flexion and dorsiflexion occur in the sagittal plane and eversion and inversion in the frontal plane, this makes the ankle joint a biaxial joint. As a biaxial joint, the ankle is capable of circumduction (a combination of the fundamental movements of a biaxial or triaxial joint).

Movement between the tarsal bones of the foot is similar to that of the carpal bones of the wrist. The bones glide over each other to produce slight movement. These gliding movements of the tarsals combined with the movements of the ankle joint result in either **supination** of the foot (inversion of the ankle and adduction of the foot) or **pronation** of the foot (eversion of the ankle and abduction of the foot). Again, the motion between the hindfoot and midfoot bones also contributes to these movements.

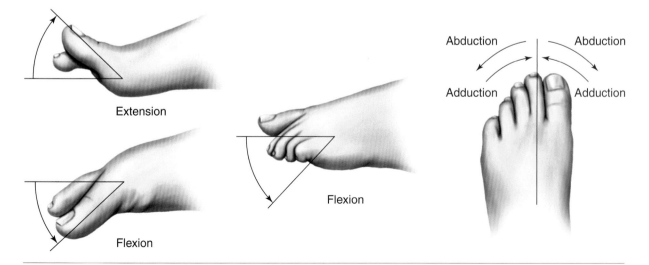

Figure 12.19   Extension and flexion of the toe joints. Abduction and adduction of the metatarsophalangeal joints.

Focus on . . . TURF TOE

Spraining the ligaments of the first (great toe) MP joint (often from hyperextension) is frequently referred to as "turf toe," depending on the mechanism of injury. Knowledge of anatomy, physics, and biomechanics, combined with the experience of engineers familiar with the effects of the interaction of footwear and the playing surface, can assist in preventing a sprain of the first MP joint.

Movements at the tarsometatarsal, metatarsophalangeal (MP), proximal interphalangeal (PIP), interphalangeal (IP), and distal interphalangeal (DIP) joints are limited to flexion and extension, except that the MP joints are also able to abduct and adduct (figure 12.19).

Muscles of the Lower Leg, Ankle, and Foot

The muscles of the leg, ankle, and foot, like those of the hand, are typically divided into the **extrinsic muscles** (those originating outside the foot and

inserting within the foot) and the **intrinsic muscles** (those originating and inserting within the foot).

Now that we've considered the movements possible in the ankle (talocrural and talocalcaneal) joint and in the MP, PIP, IP, and DIP joints, learning the actions and locations of these muscles should be easier.

Extrinsic Muscles

There are 12 extrinsic muscles of the foot and ankle that are contained in four well-defined compartments of the lower leg (figure 12.20). Four muscles are found in the **anterior compartment**,

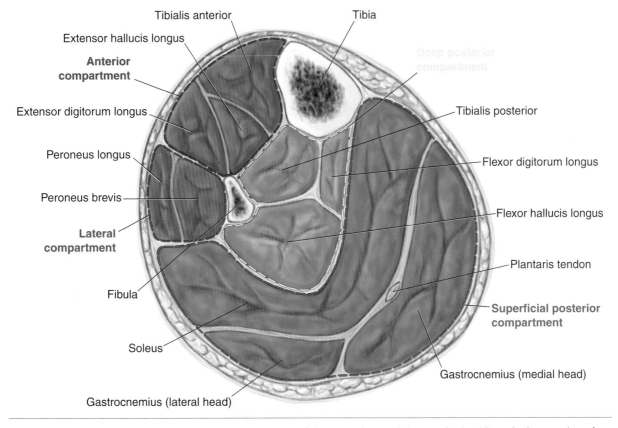

Figure 12.20   Cross-section of the four compartments of the lower leg, which contain the 12 extrinsic muscles of the ankle and foot.

two in the **lateral compartment,** three in the **superficial posterior compartment,** and three in the **deep posterior compartment.**

Anterior Compartment

The four muscles of the lower-leg anterior compartment are the tibialis anterior, the extensor digitorum longus, the extensor hallucis longus, and the peroneus tertius (figure 12.15).

> **Tibialis anterior:** The tibialis anterior originates from the upper two-thirds of the lateral side of the tibia and inserts on the medial side of the medial cuneiform and the base of the first metatarsal bone (figures 12.15 and 12.21). If we consider the origin and insertion of the tibialis anterior, what actions of the ankle joint are produced by contraction of this muscle? (Answer: dorsiflexion and inversion)

Hands on . . . Find your anterior tibial shaft and move your fingers just lateral to the tibia. The first soft tissue you palpate is your tibialis anterior muscle. Additionally, as you dorsiflex your ankle, observe the tendon of the tibialis anterior on the anterior medial surface of the ankle joint.

> **Extensor digitorum longus:** The second extrinsic muscle of the anterior compartment is the extensor digitorum longus (figure 12.15). The extensor digitorum longus originates from the lateral condyle of the tibia, the proximal three-quarters of the fibula, and the interosseous membrane and inserts on the middle and distal phalanges of the lateral four toes. From this muscle's name and its origin and insertion, what are the actions of this muscle at the ankle joint and the MP, PIP, and DIP joints? (Answer: dorsiflexion of the ankle and extension of the MP, PIP, and DIP joints of the four lateral toes)

Hands on . . . As you dorsiflex your ankle, observe the tendons of the extensor digitorum longus on the anterior surface of the ankle joint (lateral to the tibialis anterior) (figure 12.22). Also observe how the tendon splits into four tendons of insertion that should be visible on the superior (dorsal) aspect of your foot.

> **Extensor hallucis longus:** The third muscle of the lower-leg anterior compartment is the extensor hallucis longus (figure 12.15). Just as structures in the upper extremity use the term *pollicis* to mean thumb, in the lower extremity, the term *hallucis* refers to the great toe. The extensor hallucis longus

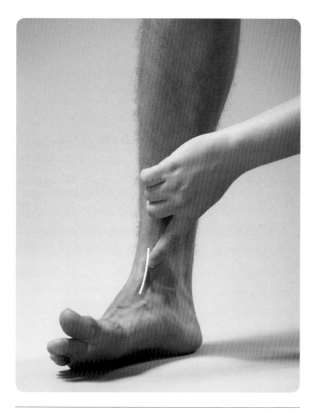

Figure 12.21 Finding the tibialis anterior.

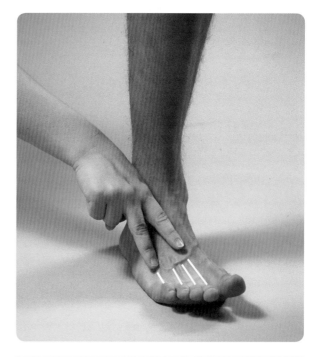

Figure 12.22 Locating the extensor digitorum longus.

originates on the middle half of the medial aspect of the fibula and the adjoining interosseous membrane and inserts on the distal phalanx of the great toe. As you observe the joints that this muscle crosses, what would you expect are its actions at the ankle and the MP and IP joints of the great toe? (Answer: dorsiflexion and inversion of the ankle and extension of the MP and IP joint of the great toe)

Hands on . . . Extend your great toe and observe the tendon of insertion of the extensor hallucis longus (figure 12.23). Tracing it from the great toe to the ankle joint, note that it quickly disappears as it moves beneath the muscle fibers of the two previously discussed anterior compartment muscles, the tibialis anterior and the extensor digitorum longus (see figure 12.15).

> **Peroneus tertius:** The fourth and final muscle of the anterior compartment of the lower leg is the peroneus tertius (the term *tertius* means third). The peroneus tertius originates from the lower third of the fibula and adjoining interosseous membrane and inserts on the superior (dorsal) surface of the base of the fifth metatarsal bone (see figure 12.15).

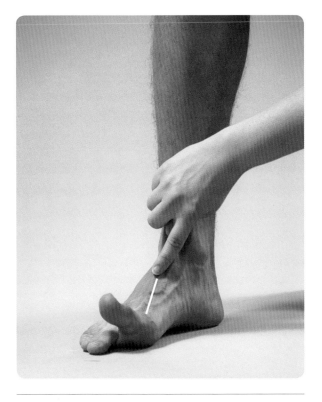

Figure 12.23   Locating the extensor hallucis longus.

Because the peroneus tertius crosses the anterior aspect of the ankle and is the most lateral of the four anterior compartment muscles, what would you expect the actions of the peroneus tertius muscle to be? (Answer: dorsiflexion and eversion of the ankle joint).

Lateral Compartment

Two muscles are contained in the lateral compartment of the lower leg: the peroneus longus and the peroneus brevis (figures 12.15 and 12.24).

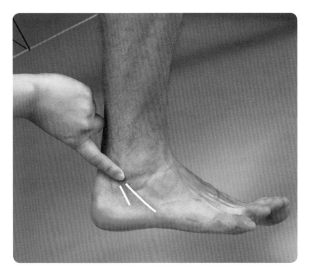

Figure 12.24   Finding the peroneus longus and peroneus brevis.

> **Peroneus longus:** The peroneus longus originates from the head and proximal two-thirds of the lateral aspect of the fibula and the lateral condyle of the tibia and inserts on the inferior lateral surfaces of the medial cuneiform and the first metatarsal bones. It was previously noted in this chapter that the cuboid bone has a concavity (groove) along its lateral border. This concavity is where the tendon of insertion of the peroneus longus passes from the lateral aspect of the foot to the inferior plantar surface.

> **Peroneus brevis:** The peroneus brevis (*brevis* means short) originates from the lower half of the lateral aspect of the fibula and inserts on the tuberosity on the base of the fifth metatarsal bone.

Note that the tendons of insertion of both the peroneus longus and peroneus brevis muscles run posterior to the lateral malleolus of the fibula. To ensure that these tendons remain in their appropriate positions while the ankle joint dorsiflexes and plantar flexes, the tendons are held in place by two structures previously discussed in this chapter: the superior and inferior peroneal retinacula (see figure 12.15). The action of both of these muscles is the same. If we consider the path that these muscles and their tendons of insertion follow, what are these actions? (Answer: plantar flexion and eversion of the ankle joint)

Superficial Posterior Compartment

The three muscles of the superficial posterior compartment of the lower leg are the gastrocnemius, the soleus, and the plantaris. Collectively, these are often referred to as the muscles of the calf (figure 12.16) and also are referred to as the **triceps surae**. Two of these muscles originate superior to the lower leg on the femur of the thigh (gastrocnemius and plantaris), whereas the third (soleus) originates on the lower leg (tibia only).

> **Gastrocnemius:** The gastrocnemius muscle is a two-headed muscle (figures 12.16 and 12.25). Its lateral head originates on the popliteal surface of the femur just medial to the lateral condyle, whereas the medial head originates on the popliteal surface of the medial condyle of the femur. Both heads combine to form the belly of the muscle, and a common tendon of insertion, known as the Achilles tendon, attaches the muscle to the posterior surface of the calcaneus. Because the gastrocnemius crosses both the knee joint and the ankle joint, extension of the knee combined

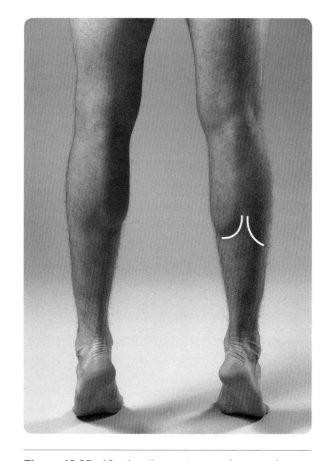

Figure 12.25   Viewing the gastrocnemius muscle.

with dorsiflexion of the ankle can contribute to an Achilles tendon strain.

> **Soleus:** The soleus (figure 12.16) originates from the head and proximal third of the posterior aspect of the fibula and the middle third of the

---

Focus on . . . **ACHILLES TENDON**

The Achilles tendon is often a source of concern in athletics. The gastrocnemius and soleus muscles are subject to strain as a result of overuse, and these strains are common in middle-aged tennis players, sports officials, and runners. These strains are often referred to as "tennis leg." We've already described the tendon of insertion of both the gastrocnemius and soleus muscles as the Achilles tendon. This tendon is often subject to strain from overuse as a result of increased activity (running, jumping), defined as an increase in frequency, intensity, duration, or any combination of these factors. Any of this can lead to inflammation of the tendon, which is poorly vascularized to begin with, making recovery more complicated. Achilles tendinitis (inflammation) can lead to a rupture of the tendon as a result of a progressive degeneration of the tissue. Additionally, a rupture of the tendon can also result from excessive force in the form of (a) direct force to, (b) forceful contraction of, or (c) forceful loading of the gastrocnemius/soleus muscles.

posterior aspect of the tibia. It inserts on the calcaneus via the Achilles tendon, the same tendon of insertion as the gastrocnemius.

Hands on . . . You can easily palpate the Achilles tendon just proximal to your calcaneus.

> **Plantaris:** The third muscle that occurs in the superficial posterior compartment is the plantaris muscle (figure 12.16). This muscle originates from the lateral linea aspera and the oblique popliteal ligament and inserts on the posterior aspect of the calcaneus (separately from the Achilles tendon). Although this short-bellied muscle lays claim to having the longest tendon of any muscle in the body, it is of minor significance in assisting other muscles of the knee and ankle joints. The plantaris crosses the knee and ankle in the same fashion as the gastrocnemius and is strained by the same mechanism. Its function is that of an assistant to the knee flexors and ankle plantar flexors. It is not a major joint mover and is analogous in function to the palmaris longus in the forearm and wrist.

Because all three muscles of the superficial posterior compartment insert on the posterior aspect of the calcaneus, their action at the ankle joint is limited to one movement. What action do all three of these muscles perform at the ankle joint? (Answer: plantar flexion of the ankle)

Deep Posterior Compartment
Three muscles, which bear names very similar to those of the anterior compartment, are found in the deep posterior compartment of the lower leg. These three muscles are the tibialis posterior, the flexor digitorum longus, and the flexor hallucis longus (see figure 12.16).

> **Tibialis posterior:** The tibialis posterior originates from the middle third of the posterior aspect of the tibia, the proximal two-thirds of the medial aspect of the fibula, and the interosseous membrane. It inserts on the inferior (plantar) surfaces of the navicular and medial cuneiform bones and the bases of the second, third, fourth, and fifth metatarsal bones (figures 12.16 and 12.26). Observe that the tendon of insertion passes posterior to the medial malleolus of the tibia: What are the likely actions of this muscle? (Answer: plantar flexion and inversion of the ankle joint)

> **Flexor digitorum longus:** The flexor digitorum longus originates from the lower two-thirds of the posterior aspect of the tibia and inserts on the bases of the distal phalanges of the four lateral toes (figures 12.16 and 12.27). Because it crosses the medial aspect of the ankle and splits into four tendons running through the plantar surface of the foot to the distal phalanges, what are the actions performed by this muscle? Don't forget to look carefully at the name of the muscle. (Answer: plantar flexion and inversion of the ankle joint and flexion of the MP, PIP, and DIP joints of the four lateral toes)

> **Flexor hallucis longus:** The flexor hallucis longus originates from the lower two-thirds of the posterior aspect of the fibula and interosseous membrane and inserts on the base of the distal phalanx of the great toe (figures 12.16 and 12.28). Because it crosses the medial aspect of the ankle and runs through the plantar surface of the foot to the distal phalanx, what are the actions performed by this muscle? Don't forget to look carefully at the name of the muscle. (Answer: plantar flexion and inversion of the ankle joint and flexion of the MP and IP joints of the great toe)

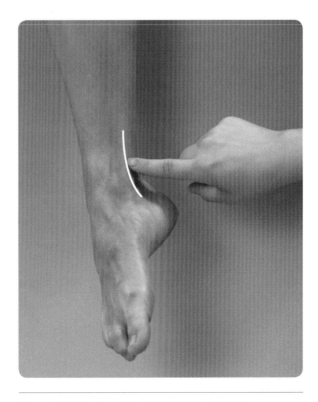

Figure 12.26   Locating the tibialis posterior.

Focus on . . . **SHIN SPLINTS**

The term *shin splints* has been called a "wastebasket term" by many because it has been used to describe a variety of conditions involving the lower leg. The term has been used to describe inflammation of the tendons of both the anterior and posterior tibialis muscles, stress fractures of the tibia and the fibula, inflammation of the periosteum of the tibia and the fibula (periostitis), and inflammation of the interosseous membrane between the fibula and the tibia. Another accepted description is that of osteoperiostitis (inflammation of the periosteum of the tibia) resulting from overuse (repetitive loading) of the posterior tibialis and soleus muscles. Causes range from improper footwear, problems of the longitudinal arch, playing surfaces, and repetitive training activities (resulting in multiple symptoms referred to as an "overuse syndrome"). Whatever the definition, whatever the cause, attention must be given to lower leg pain. "Running it out" is not the answer. Ignoring a leg injury can result in continuing inflammation that can develop an ischemia (an anemia caused by obstruction of circulation) that can lead to serious conditions such as necrosis (death of tissue) and stress fractures.

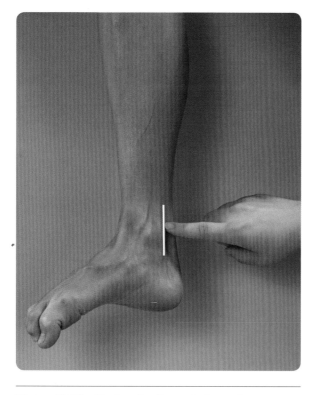

Figure 12.27   Finding the flexor digitorum longus.

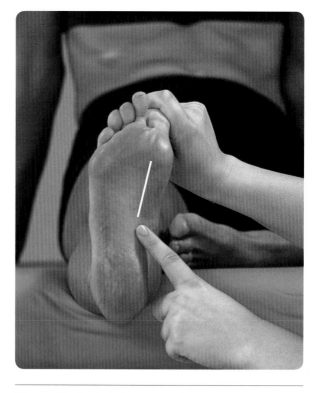

Figure 12.28   Pointing out the flexor hallucis longus.

Intrinsic Muscles of the Foot

The majority of intrinsic muscles of the foot are found on the plantar surface, where the bony and ligamentous arrangements create space to accommodate these structures. These muscles are found in four distinct layers and are very similar in arrangement to the intrinsic muscles of the hand. Note that the names of many of these muscles indicate their actions.

Superficial to all muscles on the plantar surface of the foot is a structure known as the **plantar fascia** (figure 12.29). This fibrous band originates from the calcaneus; inserts on the five MP joints;

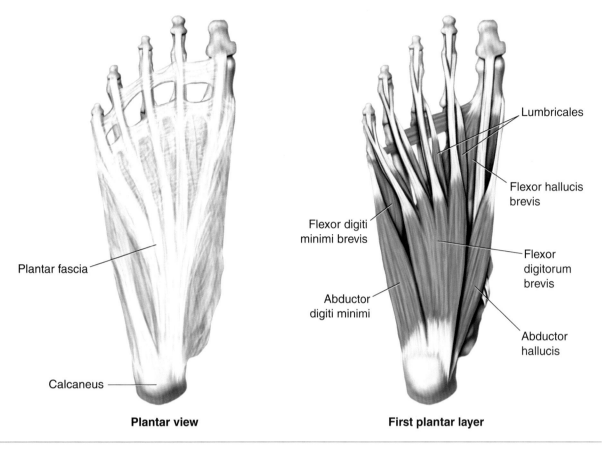

Plantar view   **First plantar layer**

Figure 12.29   The plantar fascia and muscles in the first plantar layer of the foot.

and helps to protect the muscles, blood vessels, and nerves running through the plantar surface of the foot.

First Layer

Beneath the plantar fascia, the first plantar layer contains three muscles (figure 12.29): the flexor digitorum brevis, the abductor hallucis, and the abductor digiti minimi.

> **Flexor digitorum brevis:** The flexor digitorum brevis originates from the tuberosity on the plantar surface of the calcaneus and inserts on the middle phalanges of the four lateral toes. This muscle flexes the MP and PIP joints of the four lateral toes.

> **Abductor hallucis:** The abductor hallucis originates from the medial aspect of the plantar surface of the calcaneus and inserts on the medial surface of the proximal phalanx of the great toe. The muscle abducts (moves away from the second toe) the MP joint of the great toe and also assists with flexion of the MP joint of the great toe.

Focus on . . . PLANTAR FASCIITIS

A common problem for runners is inflammation of the area of origin of the plantar fascia from repeated stress. This condition is known as *plantar fasciitis*. The point of attachment of the plantar fascia is just anterior to the tubercle of the inferior aspect of the calcaneus. Stress placed on the foot through the repetitive action of running can cause inflammation of this area. Prevention and treatment of this condition are often concerns for orthopedists and podiatrists.

> **Abductor digiti minimi:** The abductor digiti minimi originates from the tuberosity on the plantar surface of the calcaneus and inserts on the lateral aspect of the base of the proximal phalanx of the fifth toe. The muscle abducts the MP joint of the fifth (little) toe and also assists with flexion of the fifth MP joint.

Second Layer

The second plantar layer contains two muscles, quadratus plantae and the lumbricales (figure 12.30), with some unique characteristics.

> **Quadratus plantae:** The quadratus plantae muscle has two heads—one from the medial surface of the calcaneus and the other from the lateral surface of the calcaneus—which combine to attach to the tendons of the flexor digitorum longus. This muscle assists with flexion of the MP, PIP, and DIP joints of the lateral four toes.

> **Lumbricales:** The other muscle of the second plantar layer is really a group of four muscles collectively known as the lumbricales. These four muscles have no bony attachments: They originate from the tendon of the flexor digitorum longus and insert on the tendon of the extensor digitorum longus. The lumbricales flex the MP joints and extend the PIP and DIP joints of the lateral four toes.

Third Layer

The third plantar layer contains three muscles whose names indicate their functions (figure 12.30): the flexor hallucis brevis, the adductor hallucis, and the flexor digiti minimi brevis.

> **Flexor hallucis brevis:** The flexor hallucis brevis originates from the three cuneiform bones and inserts on the base of the proximal phalanx of the great toe. It flexes the MP joint of the great toe, as its name indicates.

> **Adductor hallucis:** The adductor hallucis has two heads: One is an oblique head, which originates from the bases of the second, third, and fourth metatarsals and inserts on the proximal phalanx of the great toe. The second head is a transverse head, which originates from the plantar surface of the MP capsular ligaments of the third, fourth, and fifth toes and inserts on the proximal

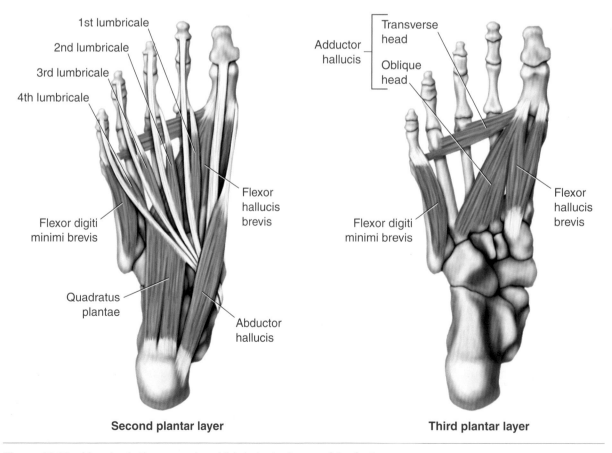

Second plantar layer

Third plantar layer

Figure 12.30 Muscles in the second and third plantar layers of the foot.

phalanx of the great toe. As its name indicates, this muscle adducts the great toe.

> **Flexor digiti minimi brevis:** The flexor digiti minimi brevis originates from the base of the fifth metatarsal bone and inserts on the lateral aspect of the base of the proximal phalanx of the fifth (little) toe. Its function, as its name indicates, is to flex the MP joint of the little toe.

Fourth Layer

The fourth layer on the plantar surface of the foot contains two groups of muscles: the dorsal interossei and plantar interossei found between the bones (interosseous) of the foot (figure 12.31).

> **Dorsal interossei:** There are four dorsal interosseous muscles. The dorsal interossei originate from the adjacent sides of all metatarsal bones and insert on the bases of the second, third, and fourth proximal phalanges. Note that the midline of the foot is considered the second toe (as opposed to the middle finger in the hand). The action of the dorsal interossei results in abduction of the third and fourth MP joints. Because there are dorsal interosseous muscles on either side of the second toe, contraction of the dorsal interossei results in no movement of the second toe.

> **Plantar interossei:** There are three plantar interosseous muscles. The plantar interossei originate from the medial surfaces of the third, fourth, and fifth metatarsal bones and insert on the medial aspect of the bases of the proximal phalanges of the third, fourth, and fifth toes. The action of the plantar interossei results in the adduction of the MP joints of the third, fourth, and fifth toes. Note again that the second toe does not move on contraction of the plantar interosseous muscles. Is this for the same reason that the second toe did not move when the dorsal interossei contracted? If not, why does the second toe not move when the plantar interosseous muscles contract?

Dorsal Intrinsic Muscles of the Foot

Although there is not as much space on the dorsal aspect of the foot as there is on the plantar aspect, there are two intrinsic muscles (extensor digitorum brevis and extensor hallucis brevis) on the dorsal aspect (figure 12.15).

> **Extensor digitorum brevis:** The extensor digitorum brevis (figure 12.32) originates from the calcaneus and inserts on the tendons of the extensor digitorum longus, which in turn insert on the second, third, and fourth toes.

> **Extensor hallucis brevis:** A fourth tendon of the extensor digitorum brevis branches to the distal phalanx of the great toe and is often called the extensor hallucis brevis muscle.

As their names indicate, these two muscles assist with extension of the MP and IP joints of the great toe and the MP, PIP, and DIP joints of the second, third, and fourth toes.

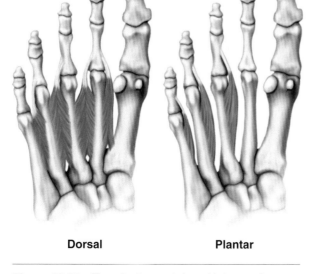

Dorsal   **Plantar**

Figure 12.31 The plantar and dorsal interossei.

Figure 12.32 Locating the extensor digitorum brevis.

REVIEW OF TERMINOLOGY

The following terms were discussed in this chapter. Define or describe each term, and where appropriate, identify the location of the named structure either on your body or in an appropriate illustration.

abductor digiti minimi
abductor hallucis
Achilles tendon
adductor hallucis
ankle joint
anterior compartment
anterior talofibular ligament
anterior talotibial ligament
anterior tibiofibular ligament
base of a metatarsal
bifurcated ligament
calcaneocuboid ligament
calcaneofibular ligament
calcaneonavicular ligament
calcaneotibial ligament
calcaneus
cuboid bone
cuboideonavicular ligament
cuneiform bones
cuneocuboid ligament
cuneonavicular ligament
deep posterior compartment
deltoid ligament
dorsal interossei
dorsal talonavicular ligament
dorsiflexion
eversion
extensor digitorum brevis
extensor digitorum longus
extensor hallucis brevis
extensor hallucis longus
extrinsic muscles
fibula
fibular notch of the tibia
flexor digiti minimi brevis
flexor digitorum brevis

flexor digitorum longus
flexor hallucis brevis
flexor hallucis longus
flexor retinaculum
forefoot
gastrocnemius
head of a metatarsal
hindfoot
intercuneiform ligaments
intermetatarsal ligaments
interosseous ligament (tibiofibular)
interosseous (collateral)
    ligaments (tarsometatarsal,
    metatarsophalangeal,
    interphalangeal)
interphalangeal ligaments
intertarsal ligaments
intrinsic muscles
inversion
lateral compartment
lateral malleolus
longitudinal arch
long plantar ligament
lumbricales
malleoli
medial malleolus
metatarsal arch
metatarsal bones
metatarsophalangeal ligaments
midfoot
navicular bone
peroneus brevis
peroneus longus
peroneus tertius
phalanges

plantar fascia
plantar flexion
plantar interossei
plantaris
posterior talofibular ligament
posterior talotibial ligament
posterior tibiofibular ligament
pronation
quadratus plantae
scaphoid
sesamoid
shaft of a metatarsal
sinus tarsi
soleus
subtalar joint
superficial posterior compartment
superior and inferior extensor
    retinacula
superior and inferior peroneal
    retinacula
supination
sustentaculum tali
talocalcaneal joint
talocalcaneal ligament
talus
tarsal bones
tarsometatarsal ligaments
tibia
tibialis anterior
tibialis posterior
tibiofibular joint (syndesmosis)
tibionavicular ligament
transverse arch
transverse metatarsal ligament
triceps surae

SUGGESTED LEARNING ACTIVITIES

1. While standing in the anatomical position, perform the following:

    a. Stand on your toes. What position does this place your ankle joints in, and what muscles produced this movement?

    b. Stand on your heels. What position does this place your ankle joints in, and what muscles produced this movement?

2. Someone you know has a sprained ankle. The physician says that this is the most common of all ankle sprains and that part of the rehabilitation routine should concentrate on strengthening the everter muscles.

    a. What ligament did this person likely sprain?

    b. What specific muscles does the physician recommend be strengthened?

3. From a standing position, go down into a baseball or softball catcher's position.

    a. Did your heels rise up off the floor?

    b. If so, why? What muscles caused the heels to rise?

4. Look carefully at the peroneal tendons on the lateral side of the ankle as the ankle dorsiflexes and plantar flexes. If, for some reason, the peroneal retinacula failed to perform their function and the peroneal tendons were allowed to move forward in front of the lateral malleolus of the fibula, how would the function of the peroneal muscles be altered?

5. While you are walking, what is the position of the ankle joint when your heel strikes the walking surface, and what is the position of the ankle joint when your foot "toes off" and no longer bears weight? How would a weakness in either the posterior compartment muscles or the anterior compartment muscles affect walking?

MULTIPLE-CHOICE QUESTIONS

1. Which of the following muscles attaches to the base of the fifth metatarsal bone?

    a. peroneus longus
    b. peroneus brevis
    c. tibialis anterior
    d. tibialis posterior

2. Which malleolus at the ankle joint is the most distal in relationship to the other?

    a. lateral
    b. medial
    c. anterior
    d. posterior

3. How many tarsal bones are found in the foot?

    a. 2
    b. 5
    c. 7
    d. 26

4. Which of the tarsal bones sits on top of (superior to) the calcaneus?

    a. scaphoid
    b. talus
    c. cuboid
    d. lateral cuneiform

5. The quadratus plantae muscle assists which of the following muscles in flexing the toes?

    a. flexor digitorum longus
    b. flexor digitorum brevis
    c. flexor hallucis longus
    d. flexor hallucis brevis

6. Pronation of the foot combines abduction of the foot and which of the following ankle joint movements?

    a. rotation
    b. circumduction
    c. inversion
    d. eversion

7. How many phalanges are found in the forefoot?

    a. 5
    b. 14
    c. 15
    d. 19

8. Supination of the foot combines adduction of the foot and which of the following ankle joint movements?

    a. rotation
    b. circumduction
    c. inversion
    d. eversion

9. Which of the following muscles is considered a muscle of the lower leg's anterior compartment?

    a. peroneus magnus
    b. peroneus longus
    c. peroneus brevis
    d. peroneus tertius

10. Which of the following muscles is not found in the superficial posterior compartment of the leg?

    a. plantaris
    b. tibialis posterior
    c. gastrocnemius
    d. soleus

11. Of the following muscles, which is not found in the deep posterior compartment of the leg?

    a. tibialis posterior
    b. flexor digitorum longus
    c. plantaris
    d. flexor hallucis longus

FILL-IN-THE-BLANK QUESTIONS

1. The subtalar joint is the joint between the talus and the _____.

2. The largest bone of the foot is the _____ _____.

3. The only function of the soleus muscle is ankle _____.

4. The gastrocnemius muscle originates on the _____.

5. The medial ligaments of the ankle joint are often referred to collectively as the _____ _____ ligament.

6. Fundamental ankle joint movements in the sagittal plane include plantar flexion and _____ _____.

7. The medial malleolus is found at the distal end of the _____.

8. The abductor hallucis muscle's function is to _____ _____ the _____ joint.

9. Eversion of the ankle joint is an attempt to move the plantar surface of the foot _____.

10. The bases of the long metatarsal bones of the foot form the _____ arch.

11. The most lateral bone in the midfoot is the _____ _____.

12. Muscles originating on the leg, crossing the ankle, and inserting on the foot are considered _____ _____ muscles.

13. The adductor hallucis muscle has two heads: the oblique and the _____.

14. The Achilles tendon consists of the soleus and _____ tendons.

15. The dorsal interosseous muscles of the foot _____ _____ the toes.

16. The anatomical name of the large ball of the foot is the _____ joint.

17. In addition to dorsiflexing the ankle joint, the tibialis anterior muscle also _____ the _____.

18. The plantar interossei of the foot _____ _____ the toes.

19. The longitudinal arch of the foot runs from the calcaneus to the _____.

20. In addition to being a major inverter of the ankle, the tibialis posterior muscle also _____ _____ the _____.

CHAPTER THIRTEEN

Nerves and Blood Vessels of the Lower Extremity

As in both the upper extremity and the spinal column and thorax, the musculature of the lower extremity is innervated by spinal nerves that are formed into plexuses. The spinal nerves that innervate the muscles of the lower extremity arise from the **lumbosacral plexus,** which is typically divided into the **lumbar plexus** (T12, L1, L2, L3, L4; figure 13.1), the **sacral plexus** (L4, L5, S1, S2, S3; figure 13.1), and the **pudendal (coccygeal) plexus** (S2, S3, S4, S5, C1, C2).

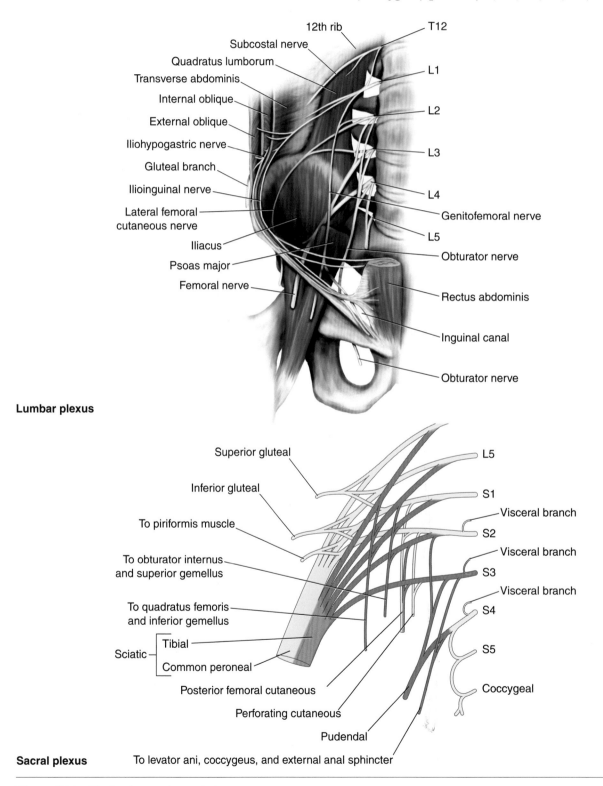

Lumbar plexus

Sacral plexus

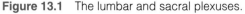

Figure 13.1   The lumbar and sacral plexuses.

The pudendal plexus, which often is considered part of the sacral plexus, innervates many of the structures of the abdominal cavity and the reproductive systems but none of the muscles of the lower extremity. Therefore, the pudendal plexus is not considered in this discussion of the lower extremity.

Nerves of the Lumbosacral Plexus

The nerves of the lumbosacral plexus are divided into three parts according to their location relative to the spinal column (superior, middle, and inferior).

Superior Portion

The most superior portion of the lumbosacral plexus, known as the lumbar plexus, consists of the **femoral, anterior femoral cutaneous, saphenous, genitofemoral, iliohypogastric, ilioinguinal,** and the **obturator nerves** and the lateral cutaneous nerve of the thigh, the **lateral femoral cutaneous** (figure 13.2).

The femoral nerve (L2, L3, L4) innervates the pectineus, the four muscles of the quadriceps

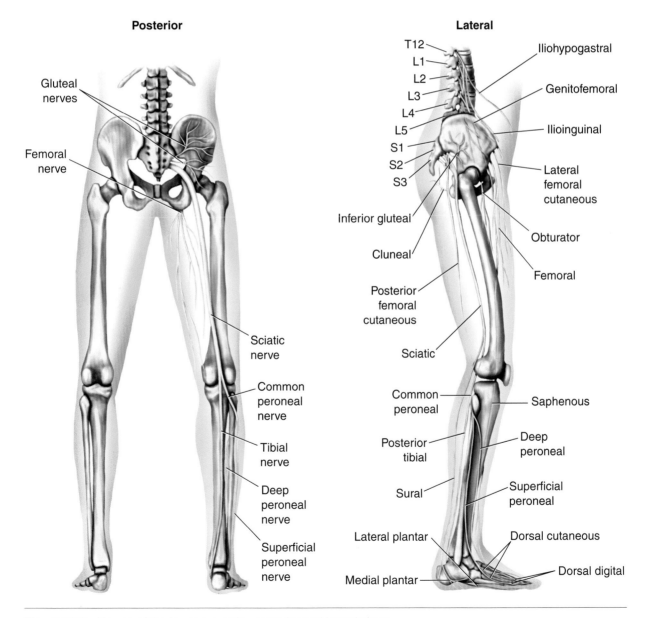

Posterior

Gluteal nerves

Femoral nerve

Sciatic nerve

Common peroneal nerve

Tibial nerve

Deep peroneal nerve

Superficial peroneal nerve

Lateral

T12
L1
L2
L3
L4
L5
S1
S2
S3

Inferior gluteal

Cluneal

Posterior femoral cutaneous

Sciatic

Common peroneal

Posterior tibial

Sural

Lateral plantar

Medial plantar

Iliohypogastral

Genitofemoral

Ilioinguinal

Lateral femoral cutaneous

Obturator

Femoral

Saphenous

Deep peroneal

Superficial peroneal

Dorsal cutaneous

Dorsal digital

Figure 13.2 Nerves of the lower extremity, posterior and lateral views.

group, and the sartorius. The cutaneous branches of the femoral nerve, the anterior femoral cutaneous nerve, and the saphenous nerve do not innervate muscles. The genitofemoral nerve (L1, L2) supplies the male and female genitalia and has a muscular branch innervating the cremaster muscle located in the male's scrotum and therefore is not considered a nerve of the lower extremity. Likewise, both the iliohypogastric (L1 and, in some individuals, also T12) and ilioinguinal (T12 and, in some individuals, also L1) nerves, although part of the lumbar plexus, have muscular branches that innervate only the abdominal muscles and are considered nerves of the trunk and not of the lower extremity. The lateral femoral cutaneous nerve (L1, L2, L3, L4) has muscular branches to the psoas major and minor muscles as well as the quadratus lumborum. Again, these muscles are typically considered muscles of the trunk and not the lower extremity. The obturator nerve (L2, L3, L4) has anterior and posterior muscular branches that innervate the adductor magnus, adductor longus, adductor brevis, gracilis, pectineus, and external obturator muscles.

Middle Portion

The middle portion of the lumbosacral plexus, known as the sacral plexus, consists of the **superior gluteal, inferior gluteal, sciatic, common peroneal, tibial,** and **sural nerves** and the **nerve to the piriformis.**

The superior and inferior gluteal nerves (L4, L5, S1, S2) innervate the three gluteal muscles, the gluteus maximus, the gluteus medius, and the gluteus minimus, as well as the tensor fascia lata muscle. The sciatic nerve (L4, L5, S1, S2, S3) is the longest and largest nerve in the body and is a combination of two nerves: the tibial (L4, L5,

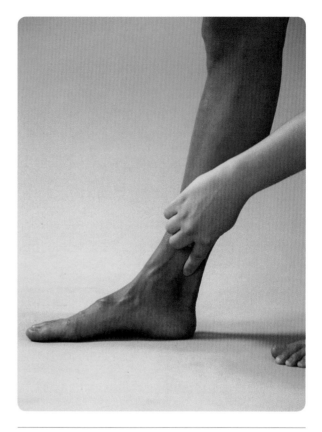

Figure 13.3 Pointing to the tibial nerve.

S1, S2, S3; figure 13.3) and the common peroneal (L4, L5, S1, S2; figure 13.4). A cutaneous branch of the common peroneal nerve, the **lateral sural nerve,** combines with the **medial sural branch of the tibial nerve** to form the sural nerve. The sural nerve and these branches do not innervate any lower-extremity muscles.

Just distal to the head of the fibula, the common peroneal nerve divides to form three terminal branches: the **deep peroneal nerve,** the **superficial peroneal nerve,** and the **recurrent tibial nerve** (figure 13.5). The deep peroneal nerve has

Focus on . . . SCIATIC NEURITIS

The sciatic nerve, because of its position and length, is often subjected to trauma in several places as a result of direct trauma, stretching, or impingement. Impingement at the lumbar or sacral disc space and muscular strains in the low back and lower extremities often cause an inflammation of the sciatic nerve called *sciatic neuritis*. Rest, anti-inflammatory agents, and, eventually, exercise are often prescribed for this condition.

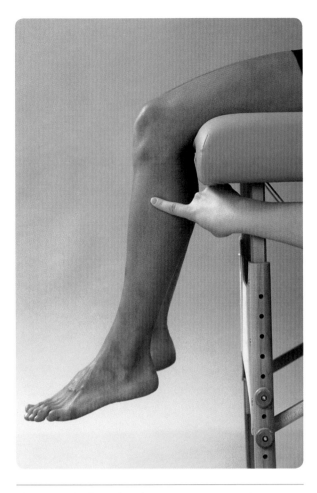

Figure 13.4   Locating the common peroneal nerve.

muscular branches innervating the four muscles of the anterior compartment in the lower leg, the extensor digitorum brevis, and the interosseous muscles of the foot. The superficial peroneal nerve innervates the peroneus longus and peroneus brevis muscles of the lateral compartment of the lower leg. The recurrent tibial nerve innervates the proximal portion of the tibialis anterior muscle. The tibial nerve has muscular branches innervating the three hamstring muscles (but not the short head of the biceps femoris). It also innervates four of the six deep external rotators of the hip joint (the superior and inferior gemelli muscles, the internal obturator muscle, and the quadratus femoris) and the muscles in both the superficial and deep posterior compartments of the lower leg. The tibial nerve has two main terminal branches: the **lateral plantar nerve** and the **medial plantar nerve.** The lateral plantar nerve has branches innervating the abductor digiti minimi, the quadratus plantae, the adductor hallucis, the lumbricales, the flexor digiti minimi brevis, and the interosseous muscles of the foot. The medial plantar nerve has branches innervating the abductor hallucis, the flexor digitorum brevis, and the lumbricales. The last nerve of the sacral plexus, the nerve to the piriformis (S2), innervates the piriformis muscle, one of the six deep external rotators of the hip joint.

Inferior Portion

The most inferior portion of the lumbosacral plexus, known as the pudendal plexus (see figure 13.1), consists of the **posterior femoral cutaneous nerve** (S2, S3), the **anococcygeal nerve** (S4, S5, C1), the **perforating cutaneous nerve** (S2, S3), and the **pudendal nerve** (S2, S3, S4). None of these nerves innervate muscles of the lower extremity, and only the pudendal nerve and anococcygeal nerve innervate muscles at all (those of the abdominal cavity and genitalia).

Focus on . . . **LOWER-LEG TRAUMA**

Trauma (e.g., contusion, overuse) to the anterior compartment of the lower leg may cause so much swelling within the compartment that pressure on the deep peroneal nerve reduces its ability to transmit an impulse to the muscles of the compartment. Loss of strength and atrophy (wasting away) of the muscle tissue can lead to complications such as a "dropped foot," resulting from an inability to properly dorsiflex the ankle. In addition to the deep peroneal nerve trauma, the function of blood vessels in the compartment may also be compromised, which can result in possible necrosis (death) of the tissues within the compartment.

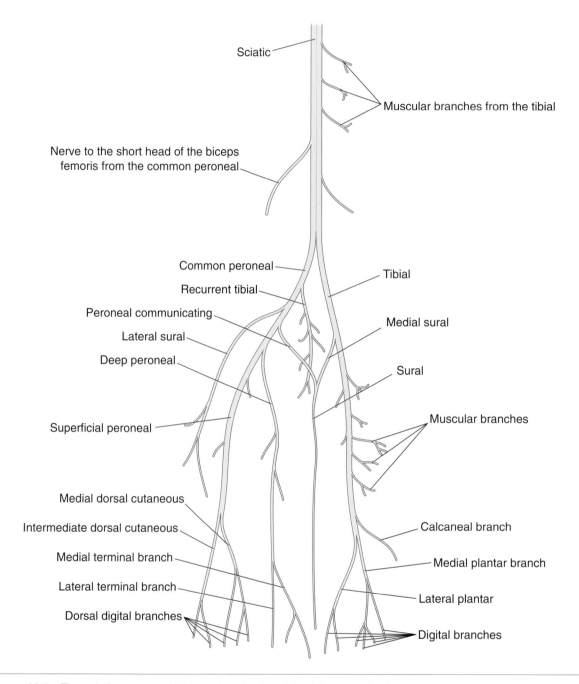

Figure 13.5   The sciatic nerve and its branches (on the right side), anterior view.

Major Arteries of the Lower Extremity

As discussed in chapter 6, the blood vessels of the body are divided into arteries, arterioles, capillaries, venules, and veins (figure 13.6).

The major arteries of the lower extremity include the **femoral,** the **popliteal,** the **anterior tibial,** and the **posterior tibial,** all with numerous branches. These arteries supply blood to the lower extremity. The femoral artery has many branches, including the **genicular arteries** that supply the adductor magnus, the gracilis, the three vastus muscles, the adductor group, the sartorius, and the vastus medialis. Another branch of the femoral artery, the **superficial circumflex iliac artery,** supplies the iliacus, tensor fascia lata, and sartorius muscles. The

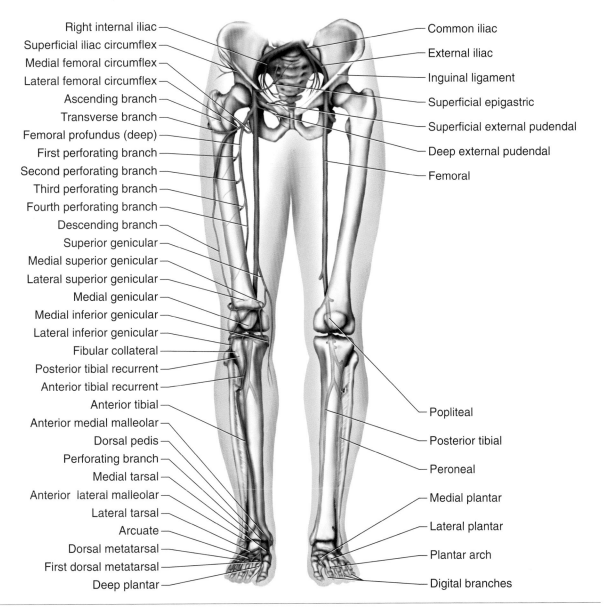

Figure 13.6 Major arteries of the lower extremity.

largest branch of the femoral artery is the **profunda artery.** It has three branches: the **lateral femoral circumflex,** the **medial circumflex,** and four **perforating arteries,** which supply the three gluteal muscles, the tensor fascia lata, the sartorius, the vastus lateralis and intermedius muscles, the rectus femoris, the adductors, the external obturator, the pectineus, and the hamstrings (figure 13.7).

Hands on . . . Apply pressure with your fingertips on your femoral triangle, and attempt to feel your pulse from the femoral artery (figure 13.8).

At the level of the knee joint, the femoral artery becomes known as the popliteal artery. The branches of the popliteal artery, the **lateral superior genicular** and **lateral inferior genicular arteries,** the **medial superior genicular** and **medial inferior genicular arteries,** the **middle genicular artery,** and its **muscular branches,** supply the vastus lateralis, adductor magnus, biceps femoris, semimembranosus, semitendinosus, popliteus, gastrocnemius, plantaris, and soleus muscles. As the popliteal artery passes the popliteal muscle, the artery divides into the

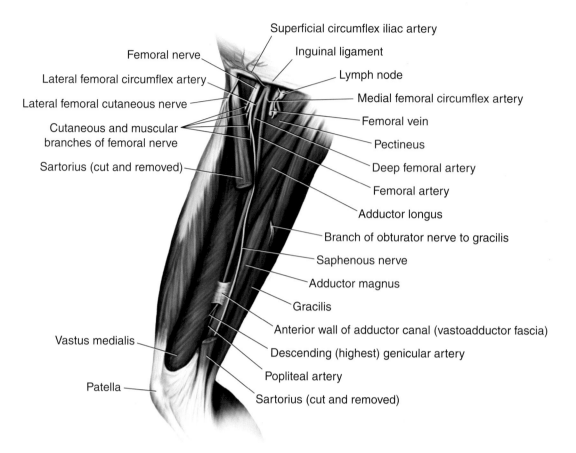

Figure 13.7 The femoral triangle.

Focus on . . . **THE FEMORAL TRIANGLE**

One of the pressure points stressed in first-aid classes for control of bleeding in the lower extremity is the femoral artery. The femoral artery is easily located within the anatomical structure known as the *femoral triangle* (see figure 13.7). The triangle is formed by the inguinal ligament (superior border), the sartorius muscle (lateral border), and the adductor longus muscle (medial border). Also passing through this area are the femoral vein and the femoral nerve.

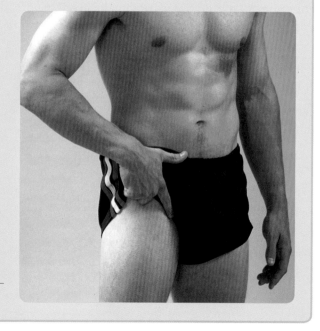

Figure 13.8 Finding the femoral artery.

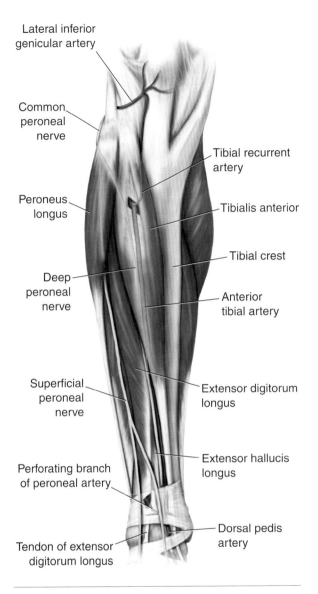

Lateral inferior
genicular artery

Common
peroneal
nerve

Peroneus
longus

Deep
peroneal
nerve

Superficial
peroneal
nerve

Perforating branch
of peroneal artery

Tendon of extensor
digitorum longus

Tibial recurrent
artery

Tibialis anterior

Tibial crest

Anterior
tibial artery

Extensor digitorum
longus

Extensor hallucis
longus

Dorsal pedis
artery

Figure 13.9   Arteries of the lower leg, anterior view.

anterior tibial and posterior tibial arteries. The anterior tibial artery has branches–**anterior** and **posterior tibial recurrent** (figures 13.6 and 13.9) and **fibular**–that supply the popliteus, extensor digitorum longus and brevis, tibialis anterior, peroneus longus, and soleus muscles. Below the ankle and into the foot, the anterior tibial artery becomes the **dorsal pedis artery** and ends on the dorsal (top) side of the foot (figures 13.6, 13.9, and 13.10). The branches of the dorsal pedis artery (**arcuate, deep plantar, first dorsal metatarsal, lateral tarsal,** and **medial tarsal**) supply the dorsal interosseous and extensor digitorum brevis muscles (figure 13.10).

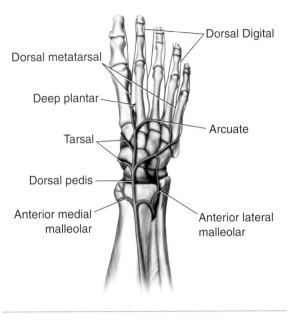

Dorsal metatarsal

Deep plantar

Tarsal

Dorsal pedis

Anterior medial
malleolar

Dorsal Digital

Arcuate

Anterior lateral
malleolar

Figure 13.10   Branches of the dorsal pedis artery.

Hands on . . . In trying to establish whether the blood supply of the foot is impeded for some reason, health care professionals attempt to palpate a dorsal pedis pulse. This pulse is normally found on the dorsal side of the foot between the proximal ends of the first and second metatarsal bones (between the extensor hallucis longus and extensor digitorum longus tendons). Use figure 13.11 to assist you in checking your dorsal pedis pulse. Not everyone can easily feel this pulse, because it has been found to be absent in 10% to 15% of all people.

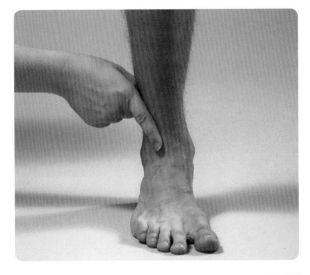

Figure 13.11   Locating the dorsal pedis artery.

The posterior tibial artery and its branches **(medial calcaneal, peroneal,** and **perforating branches)** (figure 13.12) supply the posterior and lateral lower-leg muscles, most of the muscles of the foot, and the peroneus tertius in the lower portion of the anterior lower-leg compartment. The posterior tibial artery has two terminating branches (the **lateral** and **medial plantar arteries)** that supply most of the muscles of the foot.

Hands on . . . When you are trying to determine blood flow to the foot, a more reliable pulse to take than the dorsal pedis pulse is the posterior tibial pulse. The artery is found between the flexor digitorum longus and flexor hallucis longus tendons just posterior to the medial malleolus of the tibia. Refer to figure 13.13 to palpate your posterior tibial pulse.

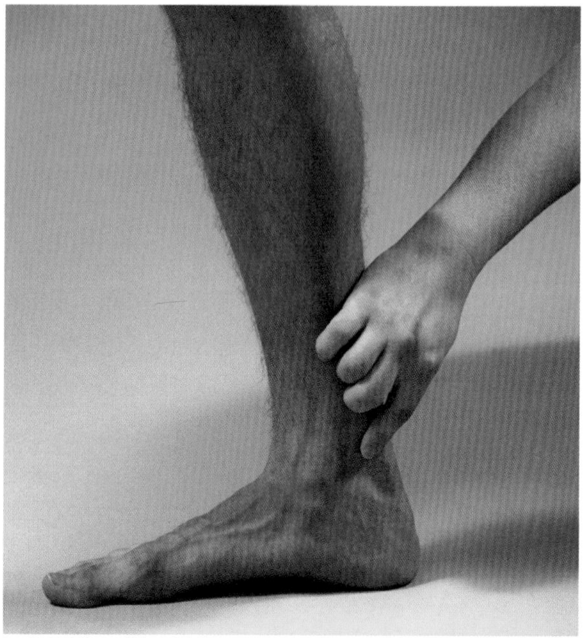

Figure 13.13   Pointing out the posterior tibial artery.

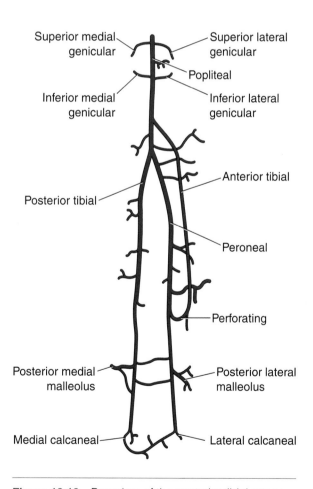

Figure 13.12   Branches of the posterior tibial artery.

Major Veins of the Lower Extremity

The veins that return blood to the heart are commonly divided into **deep veins** and **superficial veins** (figure 13.14). With a few exceptions, the deep veins have the same names as the arteries they parallel, such as the femoral, popliteal, and tibial. The superficial veins have specific names and are located near the skin. Unlike the arteries, veins do not appear exactly where one might expect, and often they are absent altogether.

The major deep veins of the lower extremity are the femoral and popliteal. The **femoral vein** drains the entire thigh and, in the area of the inguinal ligament, becomes the **external iliac vein.** The **popliteal vein,** which unites with the anterior and posterior tibial arteries, drains the structures from the foot to the lower edge of the popliteus muscle. It has tributaries known as the **anterior** and **posterior tibial veins** and the **lesser saphenous vein,** and it has branches that correspond to the anterior and posterior tibial arteries and their branches.

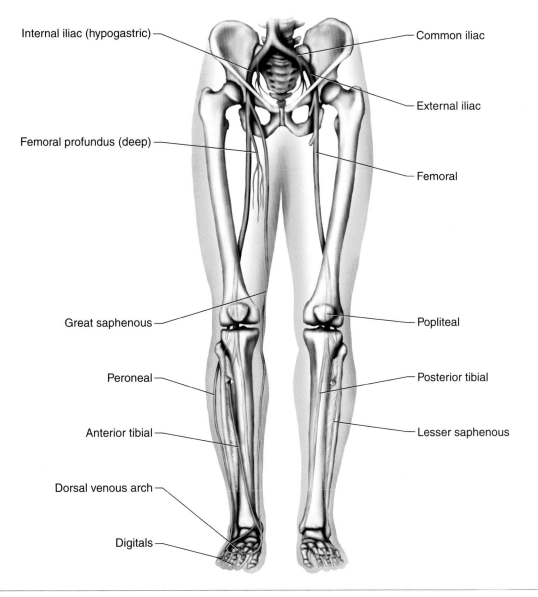

Internal iliac (hypogastric)

Common iliac

External iliac

Femoral profundus (deep)

Femoral

Great saphenous

Popliteal

Peroneal

Posterior tibial

Anterior tibial

Lesser saphenous

Dorsal venous arch

Digitals

Figure 13.14   The major veins of the lower extremity.

FOCUS ON . . . VARICOSE VEINS

Varicose veins are simply defined as distended, enlarged, swollen veins. The condition can appear in any vein in the body, but it is mentioned in this chapter because the condition most often is found in the lower extremity (thigh and lower leg). There are many causes for the condition to develop such as heredity, pregnancy, obesity, and long periods of standing. Any of these causes can lead to pressure on the veins of the lower extremity, particularly in the groin area, which can restrict blood flow return from the legs. This can cause a pooling of blood in the lower extremities that can weaken valves in the veins and actually stretch the walls of the veins. Varicose veins usually involve incompetent valves. Blood tends to pool in lower extremities weakening the valves and eventually stretching the venous walls.

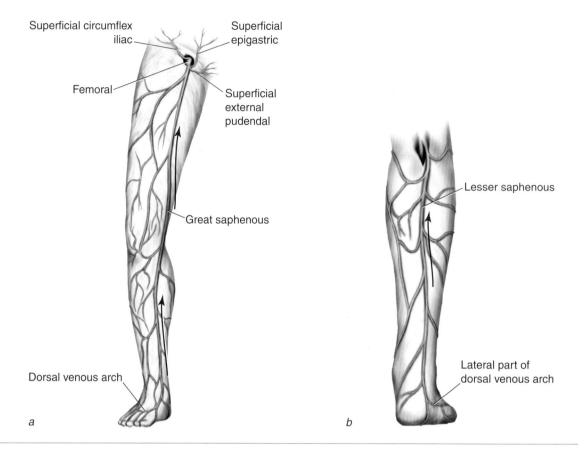

Figure 13.15   The *(a)* great and *(b)* lesser saphenous veins.

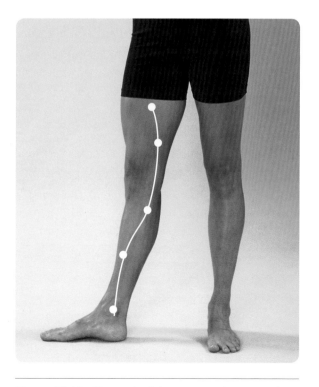

Figure 13.16   The path of the great saphenous vein.

The major superficial veins of the lower extremity include the **great saphenous** and the lesser saphenous veins (figures 13.15 and 13.16). The great saphenous vein runs between the medial aspect of the **dorsal venous arch** in the foot to the femoral triangle at the groin. This makes the great saphenous vein the longest vein in the body. Major tributaries of the great saphenous vein in the lower extremity include the **lateral** and **medial superficial femoral veins** and the **superficial circumflex iliac vein.** The lesser saphenous vein is found at the lateral aspect of the dorsal venous arch and empties into the popliteal vein in the area of the popliteal space posterior to the knee joint. The lesser saphenous vein drains the lateral and posterior aspects of the foot and lower leg. Interestingly, sections of the saphenous veins are often surgically removed and used to replace damaged or diseased sections of other blood vessels. The many peripheral branches of the saphenous vein assume the function of the removed portion.

REVIEW OF TERMINOLOGY

The following terms were discussed in this chapter. Define or describe each term, and where appropriate, identify the location of the named structure either on your body or in an appropriate illustration.

anococcygeal nerve
anterior femoral cutaneous nerve
anterior tibial artery
anterior tibial recurrent artery
anterior tibial vein
arcuate artery
common peroneal nerve
deep peroneal nerve
deep plantar artery
deep veins
dorsal pedis artery
dorsal venous arch
external iliac vein
femoral artery
femoral nerve
femoral vein
fibular artery
first dorsal metatarsal artery
genicular arteries
genitofemoral nerve
great saphenous vein
iliohypogastric nerve
ilioinguinal nerve
inferior gluteal nerve
lateral femoral circumflex artery

lateral femoral cutaneous nerve
lateral inferior genicular artery
lateral plantar artery
lateral plantar nerve
lateral superficial femoral vein
lateral superior genicular artery
lateral sural nerve
lateral tarsal artery
lesser saphenous vein
lumbar plexus
lumbosacral plexus
medial calcaneal artery
medial circumflex artery
medial inferior genicular artery
medial plantar artery
medial plantar nerve
medial superficial femoral vein
medial superior genicular artery
medial sural branch of the tibial nerve
medial tarsal artery
middle genicular artery
muscular branches of popliteal artery
nerve to the piriformis
obturator nerve
perforating arteries

perforating branches of posterior tibial
  artery
perforating cutaneous nerve
peroneal artery
popliteal artery
popliteal vein
posterior femoral cutaneous nerve
posterior tibial artery
posterior tibial recurrent artery
posterior tibial vein
profunda artery
pudendal nerve
pudendal (coccygeal) plexus
recurrent tibial nerve
sacral plexus
saphenous nerve
sciatic nerve
superficial circumflex iliac artery
superficial circumflex iliac vein
superficial peroneal nerve
superficial veins
superior gluteal nerve
sural nerve
tibial nerve

SUGGESTED LEARNING ACTIVITIES

1. Locate the anatomical area identified as the femoral triangle.

   a. Identify the three structures defining the triangle.

   b. Apply pressure to this area until you feel a pulse. What structure is providing this pulse? In typical first-aid courses, what is this area designated in terms of controlling bleeding?

2. Apply pressure with your fingers on the dorsal (top) side of the foot. Depending on the soft-tissue structures present, a pulse may be felt. If so, what structure is creating that pulse?

3. Determine whether a pulse can be established posterior to the medial malleolus of the tibia (between the medial malleolus and the Achilles tendon). If so, what structure is creating that pulse?

MULTIPLE-CHOICE QUESTIONS

1. The longest vein in the body is the

   a. femoral
   b. popliteal
   c. great saphenous
   d. tibial

2. The longest nerve in the body is the

   a. femoral
   b. popliteal
   c. sciatic
   d. common peroneal

(continued)

MULTIPLE-CHOICE QUESTIONS   (continued)

3. The dorsal venous arch drains blood from the
   a. toes
   b. anterior tibial compartment
   c. ankle
   d. posterior tibial compartment

4. The anterior and posterior tibial veins drain into the
   a. great saphenous vein
   b. lesser saphenous vein
   c. femoral vein
   d. popliteal vein

FILL-IN-THE-BLANK QUESTIONS

1. The major nerve innervating the muscles of the anterior tibial compartment is the _____ _____ nerve.

2. The _____ branch of the posterior tibial artery provides the blood supply to the peroneal muscles of the lateral compartment of the lower leg.

3. The medial adductors of the hip joint are innervated by the _____ nerve.

Joint	Type	Bones	Ligaments	Movements
Hip	Ball and socket	Ilium, ischium pubic, and femur	• Capsule • Acetabular labrium/(glenoid lip) • Iliofemoral (Y) • Pubofemoral • Transverse acetabular • Ligamentum capitis femoris (teres)	Flexion, extension, abduction, adduction, internal (or medial or inward) rotation, external (or lateral or outward) rotation
Knee				
Tibiofemoral	Modified hinge	Femur and tibia	• Capsule • Medial (tibial) collateral • Lateral (fibular) collateral • Anterior cruciate • Posterior cruciate • Oblique popliteal • Arcuate (popliteal) • Wrisberg • Coronary • Transverse • Patellar (considered a continuation of the quadriceps tendon)	Flexion, extension (lower leg internally rotates on knee flexion and externally rotates on knee extension in non-weight bearing movements)
Patellofemoral	Plane	Patella and femur	Patellar	Gliding
Lower leg				
Proximal tibiofibular	Diarthrodial (plane)	Tibia and fibula	• Capsule • Anterior of head of fibula • Posterior of head of fibula	Slight gliding
Middle tibiofibular	Synarthrodial	Tibia and fibula	Interosseous	None
Distal tibiofibular	Syndesmodial	Tibia and fibula	Interosseous • Anterior tibiofibular • Posterior tibiofibular Inferior transverse (anterior to the posterior tibiofibular)	Slight "give" on ankle dorsiflexion
Ankle				
Talotibial (talocrural)	Hinge	Tibia and talus	• Capsule • Deltoid • Tibionavicular • Calcaneotibial • Anterior talotibial • Posterior talotibial	Dorsiflexion and plantar flexion
Talofibular	Hinge	Talus and fibula	• Capsule • Calcaneofibular • Anterior talofibular • Posterior talofibular	Dorsiflexion and plantar flexion
Intertarsal joints				
Subtalar[a] (talocalcaneal)	Arthrodial	Talus and calcaneus	• Capsule • Anterior talocalcaneal • Posterior talocalcaneal • Lateral talocalcaneal • Medial talocalcaneal • Interosseous talocalcaneal	Gliding
Transverse tarsal[a]	Arthrodial	Talus, navicular, calcaneus, and cuboid	• Bifurcated (the calcaneonavicular part) • Plantar calcaneonavicular (or spring ligament)	Gliding

(continued)

Joint	Type	Bones	Ligaments	Movements
Intertarsal joints *(continued)*				
Talocalcaneo-navicular	Arthrodial	Talus, navicular, and calcaneus	• Capsule • Dorsal talonavicular • Plantar calcaneonavicular	Gliding
Calcaneocuboid	Arthrodial	Calcaneus and cuboid	• Capsule • Plantar calcaneocuboid (short plantar) • Dorsal calcaneocuboid • Bifurcated • Long plantar	Gliding
Cuneonavicular	Arthrodial	3 cuneiforms navicular	• Dorsal cuneonaviculars • Plantar cuneonaviculars	Gliding
Cuboideonavicular	Arthrodial	Cuboid navicular	• Dorsal cuboideonavicular • Plantar cuboideonavicular • Interosseous cuboideonavicular	Gliding
Intercuneiform and cuneocuboid	Arthrodial	3 cuneiforms cuboid	• Dorsal intercuneiforms • Plantar intercuneiforms • Interosseous intercuneiforms	Gliding
Tarsal-metatarsal-phalangeal joints				
Tarsometatarsals	Arthrodial (plane)	3 cuneiforms cuboid and base of metatarsals	• Capsules • Plantar tarsometatarsal and dorsal tarsometatarsal • Interosseous collaterals	Gliding
Intermetatarsals	Arthrodial (plane)	Bases of metatarsals	• Dorsal capsules • Plantar capsules • Interosseous collaterals	Gliding
Metatarsophalangeal (5)	Condyloid	Heads of metatarsals and bases of proximal phalanges	• Dorsal capsules • Plantar capsules • Interosseous collaterals • Transverse metatarsal	Flexion, extension, abduction, and adduction
Interphalangeals (4 proximal interphalangeal, 4 distal interphalangeal, and 1 interphalangeal)	Hinge	Proximal, middle, and distal phalanges	• Dorsal capsules • Plantar capsules • Interosseous collaterals	Flexion and extension

[a] The combined movements of the subtalar and transverse tarsal joints produce inversion (supination, adduction, and plantar flexion) and eversion (pronation, abduction, dorsiflexion) of the ankle joint.

PART IV SUMMARY TABLE Lower-Extremity Muscles, Nerves, and Blood Supply

Muscle	Origin	Insertion	Action	Nerve	Blood supply
Iliopsoas					
Hip, anterior					
Psoas major	L1–L5 transverse processes, T12 and L1–L5 intervertebral discs and bodies	Lesser trochanter of femur	Flexion and rotation of spinal column; flexion, adduction, and external rotation of hip	Femoral	External iliac, internal iliac, and lumbar
Psoas minor	T12 and L1 bodies and intervertebral disc	Iliopectineal line	Flexion of pelvis	Lumbar	Lumbar
Iliacus	Iliac fossa, iliolumbar, and sacroiliac ligaments	Just distal to lesser trochanter of femur	Flexion, adduction, external rotation of hip	Femoral	External iliac and hypogastric
Sartorius	Anterosuperior iliac spine	Inferior to medial condyle of tibia	Flexion, abduction, and external rotation of hip; flexion of knee; internal rotation of lower leg	Femoral	Profunda and femoral
Rectus femoris	Anteroinferior iliac spine and area superior to the acetabulum	Patella and tibial tuberosity	Flexion of hip, extension of knee	Femoral	Profunda
Tensor fascia lata	Outer lip of iliac crest and between anterosuperior and anteroinferior iliac spines	Greater trochanter of femur, and (as iliotibial band) inferior and anterior to lateral condyle of tibia	Flexion, abduction, and rotation of hip	Superior gluteal	Lateral circumflex
Pectineus	Iliopectineal line and superior aspect of pubic bone	Pectineal line on femur	Flexion, adduction, and external rotation of hip	Femoral	Femoral and femoral circumflex
Hip, lateral					
Gluteus medius	Between iliac crest and anterior and posterior gluteal lines	Posterolateral edge of greater trochanter	Abduction of hip. Anterior portion: flexion and internal rotation of hip; posterior portion: extension and external rotation of hip	Superior gluteal	Superior gluteal
Gluteus minimus	Anterior and inferior gluteal lines of ilium	Anterior edge of greater trochanter	Abduction and internal rotation of hip. Anterior portion: flexion of hip; posterior portion: extension of hip	Superior gluteal	Superior gluteal

(continued)

PART IV SUMMARY TABLE *(continued)*

Muscle	Origin	Insertion	Action	Nerve	Blood supply
Biceps femoris		**Hip, posterior**			
(Long head)	Ischial tuberosity	Head of fibula and lateral tibial condyle	Extension, adduction, and external rotation of hip; flexion of knee	Tibial	Profunda branch of femoral
(Short head)	Linea aspera	Head of fibula and lateral tibial condyle	Flexion of knee, external rotation of lower leg	Peroneal	Profunda branch of femoral
Semitendinosus	Ischial tuberosity	Inferior to medial condyle of tibia	Extension, adduction, and internal rotation of hip; flexion of knee; internal rotation of lower leg	Tibial	Profunda branch of femoral and popliteal
Semimembranosus	Ischial tuberosity	Medial tibial condyle	Extension, adduction, and internal rotation of hip; flexion of knee; internal rotation of lower leg	Tibial	Profunda branch of femoral and popliteal
Gluteus maximus	Sacrum and coccyx, posterior gluteal line, and iliac crest	Gluteal line of femur and (as iliotibial band) inferior and anterior to lateral condyle of tibia	Extension, adduction, and external rotation of hip; abduction of flexed hip	Inferior gluteal	Superior and inferior gluteals, medial circumflex
Piriformis	Superior 2/3 of sacrum and greater sciatic notch	Greater trochanter	External rotation, abduction, and extension of hip	1st and 2nd sacral	Superior and inferior gluteals
Superior gemellus	Ischial spine	Greater trochanter	External rotation of hip	1st and 2nd sacral	Inferior gluteal and obturator
Internal obturator	Inner rim of obturator foramen	Greater trochanter	External rotation of hip	1st and 2nd sacral	Inferior gluteal and obturator
Inferior gemellus	Ischial tuberosity	Greater trochanter	External rotation of hip	Lumbosacral to quadratus femoris	Inferior gluteal and obturator
External obturator	Outer aspect of pubic and ischial bones at the obturator foramen	Trochanteric fossa of femur	External rotation of hip	Obturator	Inferior gluteal and obturator
Quadratus femoris	Ischial tuberosity	Intertrochanteric crest of femur	External rotation, adduction, and extension of hip	Branch from lumbosacral plexus	Inferior gluteal and medial femoral circumflex

Hip, medial					
Adductor longus	Pubic bone	Middle 1/3 of linea aspera	Adduction, flexion, and external rotation of hip	Obturator	Femoral
Adductor brevis	Pubic bone	Proximal 1/3 of linea aspera	Adduction, flexion, and external rotation of hip	Obturator	Femoral
Adductor magnus	Pubic bone and ischial tuberosity	Linea aspera and adductor tubercle	Anterior portion: adduction, flexion, and external rotation of hip; posterior portion: adduction, extension, and internal rotation of hip	Obturator and sciatic	Femoral
Gracilis	Anterior inferior aspect of pubic symphysis and inferior pubic bone	Distal to medial tibial condyle	Adduction, flexion, and external rotation of hip; flexion of knee; internal rotation of lower leg	Obturator	Femoral
Knee, anterior					
Sartorius	*See:* Hip				
Quadriceps femoris					
Rectus femoris	*See:* Hip				
Vastus lateralis	Greater trochanter; intertrochanteric line, proximal to lateral lip of linea aspera	Lateral patella, anterior aspect of lateral tibial condyle, and rectus femoris tendon	Extension of knee	Femoral	Lateral circumflex
Vastus intermedius	Proximal 2/3 of anterior of femur	Inferior aspect of patella and tendons of vastus lateralis and medialis	Extension of knee	Femoral	Lateral circumflex and profunda
Vastus medialis	Intertrochanteric line and linea aspera	Medial tibial condyle, medial patella, and medial aspect of rectus femoris tendon	Extension of knee	Femoral	Femoral
Knee, posterior					
Hamstrings					
Biceps femoris	*See:* Hip				
Semimembranosus	*See:* Hip				
Semitendinosus	*See:* Hip				
Gracilis	*See:* Hip				

(continued)

Muscle	Origin	Insertion	Action	Nerve	Blood supply
Knee, posterior (continued)					
Popliteus	Lateral femoral condyle	Proximal tibia at popliteal line	Flexion of knee, internal rotation of lower leg	Tibial	Popliteal
Gastrocnemius					
(Lateral head)	Popliteal area medial to lateral femoral condyle	Calcaneus	Flexion of knee, plantar flexion of ankle	Tibial	Posterior tibial and peroneal
(Medial head)	Medial condyle of femur	Calcaneus	Flexion of knee, plantar flexion of ankle	Tibial	Posterior tibial and peroneal
Plantaris	Lateral linea aspera and oblique popliteal ligament	Calcaneus	Flexion of knee, plantar flexion of ankle	Tibial	Popliteal
Ankle and foot (extrinsic), anterior					
Tibialis anterior	Lateral condyle of tibia and proximal 1/2 of lateral aspect of tibia	1st cuneiform and base of 1st metatarsal	Dorsiflexion of ankle, inversion of foot	Deep peroneal	Anterior tibial
Extensor digitorum longus	Lateral condyle of tibia, proximal 3/4 fibula, and interosseous membrane	Phalanges of toes 2–5	Dorsiflexion and eversion of ankle, extension of toes 2–5	Deep peroneal	Anterior tibial
Extensor hallucis longus	Middle 1/2 of fibula and interosseous membrane	Base of distal phalanx of great toe	Dorsiflexion and inversion of ankle, extension of great toe	Deep peroneal	Anterior tibial
Peroneus tertius	Distal 1/3 of fibula and interosseous membrane	Dorsal aspect of base of 5th metatarsal	Dorsiflexion and eversion of ankle	Deep peroneal	Anterior tibial and peroneal
Ankle and foot (extrinsic), lateral					
Peroneus longus	Lateral condyle of tibia, head and proximal 2/3 of lateral fibula	Lateral aspect of base of 1st metatarsal and medial cuneiform	Eversion and plantar flexion of ankle	Superficial peroneal	Peroneal
Peroneus brevis	Distal 1/2 of fibula	5th metatarsal tuberosity	Eversion and plantar flexion of ankle	Superficial peroneal	Peroneal
Ankle and foot (extrinsic), posterior					
Gastrocnemius	*See:* Knee				
Soleus	Head and proximal 1/3 of fibula, middle 1/3 of tibia	Calcaneus	Plantar flexion of ankle	Tibial	Posterior tibial and peroneal

Muscle	Origin	Insertion	Action	Nerve	Artery
Plantaris	*See:* Knee				
Tibialis posterior	Middle 1/3 of posterior tibia, head and proximal 2/3 of medial aspect of fibula, and interosseous membrane	Navicular, medial cuneiform, and 2–5 metatarsals	Plantar flexion and inversion of ankle, adduction of foot	Tibial	Posterior tibial
Flexor digitorum longus	Posterior tibia: popliteal line to within a few inches of distal end of tibia	Base of distal phalanges of toes 2–5	Plantar flexion and inversion of ankle, adduction of foot; flexion of toes 2–5	Tibial	Posterior tibial
Flexor hallucis longus	Lower 2/3 of posterior fibula and interosseous membrane	Distal phalanx of great toe	Plantar flexion and inversion of ankle, flexion of great toe	Tibial	Posterior tibial
Ankle and foot (intrinsic), 1st plantar layer					
Flexor digitorum brevis	Calcaneal tuberosity	Sides of middle phalanges of toes 2–5	Flexion of toes 2–5	Medial plantar	Posterior tibial
Abductor hallucis	Medial aspect of calcaneus	Base of proximal phalanx of great toe	Abduction and flexion of great toe	Medial plantar	Posterior tibial
Abductor digiti minimi	Calcaneal tuberosity	Lateral aspect of base of proximal phalanx of 5th toe	Abduction and flexion of 5th toe	Lateral plantar	Posterior tibial
Ankle and foot (intrinsic), 2nd plantar layer					
Quadratus plantae					
(Medial head)	Medial aspect of calcaneus	Flexor digitorum longus tendon	Assists flexion of toes 2–5	Lateral plantar	Posterior tibial
(Lateral head)	Lateral aspect of calcaneus	Flexor digitorum longus tendon	Assists flexion of toes 2–5	Lateral plantar	Posterior tibial
Lumbricales	Flexor digitorum longus tendon	Extensor digitorum longus tendon	Flexion of PIP joints (toes 2–5), extension of DIP joints (toes 2–5)	Lateral and medial plantar	Plantar
Ankle and foot (intrinsic), 3rd plantar layer					
Flexor hallucis brevis	1st, 2nd, and 3rd cuneiforms	Base of proximal phalanx of great toe	Flexion of MP joint of great toe	Medial plantar	Posterior tibial
Adductor hallucis					
(Transverse head)	Plantar surface of MP ligament of toes 3–5	Proximal phalanx of great toe	Adduction of great toe	Lateral plantar	Posterior tibial
(Oblique head)	Bases of metatarsals 2–4	Proximal phalanx of great toe	Adduction of great toe	Lateral plantar	Posterior tibial
Flexor digiti minimi brevis	Base of 5th metatarsal	Lateral aspect of base of proximal phalanx of 5th toe	Flexion of 5th toe	Lateral plantar	Posterior tibial

(continued)

Muscle	Origin	Insertion	Action	Nerve	Blood supply
Ankle and foot (intrinsic), 4th plantar layer					
Dorsal interossei (4)	Adjacent aspects of all metatarsals	Base of proximal phalanges of toes 2–4	Abduction of toes 3 and 4	Lateral plantar	Anterior tibial
Plantar interossei (3)	Inferior medial aspect of metatarsals 3–5	Medial aspect of base of proximal phalanges of toes 3–5 and extensor digitorum longus tendon	Adduction of toes 3–5	Lateral plantar	Deep and lateral plantars
Ankle and foot (intrinsic), dorsal					
Extensor digitorum brevis	Calcaneus	Long extensor tendons of toes 2–5	Extension of toes 2–5	Deep peroneal	Anterior tibial and peroneal
Extensor hallucis brevis	Calcaneus	Distal phalanx of great toe	Extension and adduction of great toe	Deep peroneal	Anterior tibial

PIP = proximal interphalangeal; DIP = distal interphalangeal; MP = metatarsophalangeal.

Answers to End-of-Chapter Questions

CHAPTER ONE

Multiple-Choice Questions

1. c
2. a
3. b
4. a
5. c
6. c

Fill-in-the-Blank Questions

1. sutured
2. bursa
3. muscle fibers
4. toward
5. away from

CHAPTER TWO

Multiple-Choice Questions

1. b
2. d
3. b
4. b

Fill-in-the-Blank Questions

1. flexion
2. a plane
3. sagittal horizontal
4. proximal

CHAPTER THREE

Multiple-Choice Questions

1. d
2. d
3. a
4. c
5. a
6. a
7. d
8. d
9. d
10. b
11. b
12. c
13. a
14. c
15. d
16. d
17. b
18. a
19. a
20. d
21. b
22. b
23. d
24. d
25. d
26. a
27. a

Fill-in-the-Blank Questions

1. coracoclavicular
2. trapezius
3. subscapularis
4. supraspinatus
5. acromion process
6. coracobrachialis
7. scapula
8. abduction
9. clavicular
10. scapula
11. latissimus dorsi

CHAPTER FOUR

Multiple-Choice Questions

1. c
2. c
3. b
4. d
5. b
6. a
7. b
8. b
9. d
10. a
11. c
12. a

Fill-in-the-Blank Questions

1. humerus
2. annular
3. ulna
4. elbow flexion
5. trochlea
6. straight line
7. pronation
8. triceps brachii
9. pronator quadratus

CHAPTER FIVE

Multiple-Choice Questions

1. a
2. b
3. a
4. b
5. a
6. b
7. a
8. c
9. b
10. c
11. a
12. c
13. c
14. a
15. d

Fill-in-the-Blank Questions

1. palmaris longus
2. midcarpal
3. flexors
4. thumb
5. radiocarpal
6. adduction
7. index
8. scaphoid
9. opposition
10. scaphoid
11. flexor retinaculum
12. collateral
13. carpometacarpal
14. scaphoid

CHAPTER SIX

Multiple-Choice Questions

1. b
2. c
3. c
4. a

Fill-in-the-Blank Questions

1. median cubital
2. dorsal scapular
3. arteries

CHAPTER SEVEN

Multiple-Choice Questions

1. c
2. d
3. b
4. a
5. b
6. c
7. b
8. d
9. d
10. b
11. b
12. d
13. d
14. b
15. a
16. a

Fill-in-the-Blank Questions

1. lumbar
2. thoracic
3. posterior longitudinal
4. atlas
5. intervertebral disc
6. pubic symphysis
7. ischium
8. lordosis
9. kyphosis
10. scoliosis

CHAPTER EIGHT

Multiple-Choice Questions

1. b
2. b
3. a

Fill-in-the-Blank Questions

1. sternum
2. manubrium
3. diaphragm

CHAPTER NINE

Multiple-Choice Questions

1. a
2. b

Fill-in-the-Blank Questions

1. phrenic
2. internal; common iliac

CHAPTER TEN

Multiple-Choice Questions

1. c
2. a
3. b
4. d
5. b
6. b
7. b
8. d
9. d
10. b

Fill-in-the-Blank Questions

1. hip joint
2. iliofemoral
3. gluteus maximus
4. ischium
5. thoracic and lumbar spine
6. posterior

CHAPTER ELEVEN

Multiple-Choice Questions

1. a
2. d
3. b
4. b
5. c
6. a
7. d
8. b
9. c
10. d
11. a
12. b
13. c
14. b

Fill-in-the-Blank Questions

1. medial condyle
2. tibia
3. transverse
4. femur
5. anterior cruciate
6. knee
7. gluteus maximus

CHAPTER TWELVE

Multiple-Choice Questions

1. b
2. a
3. c
4. b
5. a
6. d
7. b
8. c
9. d
10. b
11. c

Fill-in-the-Blank Questions

1. calcaneus
2. calcaneus
3. plantar flexion
4. femur
5. deltoid
6. dorsiflexion
7. tibia
8. abduct; first MP
9. laterally
10. transverse
11. cuboid
12. extrinsic
13. transverse
14. gastrocnemius
15. abduct
16. first MP
17. inverts; ankle
18. adduct
19. metatarsal heads
20. plantar flexes; ankle

CHAPTER THIRTEEN

Multiple-Choice Questions

1. c
2. c
3. a
4. d

Fill-in-the-Blank Questions

1. deep peroneal
2. peroneal
3. obturator

Suggested Readings

Anderson, P.D. 1990. *Human anatomy and physiology coloring workbook and study guide.* Boston: Jones & Bartlett.

Backhouse, K.M., and R.T. Hutchings. 1986. *Color atlas of surface anatomy.* Baltimore: Williams & Wilkins.

Gray, H. 1918/2000. *Anatomy of the human body.* Philadelphia: Lea & Febiger (1918); Bartleby.com (2000).

Lindsay, D.T. 1996. *Functional human anatomy.* St. Louis: Mosby.

Marieb, E. 1991. *Human anatomy and physiology.* Redwood City, CA: Benjamin/Cummings.

Netter, F.H. 1998. *Atlas of human anatomy.* Carlstadt, NJ: ICON Learning Systems.

Olson, T.R. 1996. *A.D.A.M. student atlas of anatomy.* Baltimore: Williams & Wilkins.

Pansky, B. 1996. *Review of gross anatomy.* New York: McGraw-Hill.

Van De Graaff, K.M., and S.I. Fox. 1995. *Concepts of human anatomy and physiology.* Dubuque, IA: Brown.

Index

Note: Page numbers followed by an italicized f or t refer to the figure or table on that page, respectively.

About the Author

Robert S. Behnke, HSD, is uniquely qualified to write an introductory anatomy text that is preeminently user friendly to students. Retired after 39 years of teaching anatomy, kinesiology, physical education, and athletic training courses, Dr. Behnke has been honored on several occasions for excellence in teaching—including receiving the Educator of the Year award from the National Athletic Trainers' Association. During his 11-year tenure as chair of the NATA Professional Education Committee, he initiated the petition to the American Medical Association that led to the national accreditation process for entry-level athletic training education programs.

Most of Dr. Behnke's career was spent at Indiana State University, where he was a full professor of physical education and athletic training and director of undergraduate and graduate athletic training programs. Earlier in his career he spent five years teaching at the secondary level. He had extensive experience as an athletic trainer, both in high schools and at the university level, serving as head athletic trainer at Illinois State University from 1966 to 1969 and the University of Illinois from 1969 to 1973. He was an athletic trainer at the 1983 World University Games and for the U.S. men's Olympic basketball trials in 1984. During sabbaticals in 1982 and 1989, he served as an athletic trainer for boxing, men's field hockey, team handball, ice skating, roller hockey, gymnastics, judo, and cycling at the United States Olympic Training Center. He has been in demand as a speaker and as an athletic trainer throughout the United States and internationally.

These broad experiences enabled Dr. Behnke to understand the needs of undergraduate students—and to develop an unparalleled grasp of which pedagogical approaches work and which do not. *Kinetic Anatomy* is the culmination of his unique understanding; it should be a staple in undergraduate courses for years to come.

Essentials of Interactive Functional Anatomy

Minimum System Requirements

PC

- Windows® 98/2000/ME/XP
- Pentium® processor or higher
- At least 32 MB RAM
- Monitor set to 800 x 600 or greater
- High-color display

Mac

- Power Mac®
- System 8.6 /9/OSX
- At least 64 MB RAM
- Monitor set to 800 x 600 or greater
- Monitor set to thousands of colors

How to Use This Program

PC

The program should launch automatically when the CD is inserted in the CD-ROM drive of your computer. Choose the Install button for IFA Essentials. If you don't already have QuickTime 6 installed on your computer, you should also install that. If the CD does not auto-launch, go to My Computer and double-click the HK_IFA_Ess icon. Install as stated above.

Mac

Insert the CD into the CD-ROM drive of your computer, then double-click the CD-ROM icon. Double-click the IFA Essentials for Power Mac folder, then the IFA Essentials icon. This will launch the program. If you don't already have

QuickTime 6 installed on your computer, you should also install that using the QT installer in the IFA Essentials for Power Mac folder.

Quick Start Instructions

Use the Anatomy tab to view the 3D model and click on any anatomical structure to display the relevant text. Click on the red text for hot links to additional and relevant information about the chosen anatomical component. The History function provides access to previous text articles. Use the blue arrow buttons located in the upper right-hand corner of the text interface to move sequentially through the text.

To maximize the model interface, place the cursor over the model/text interface until the double-arrow sign appears, left-click, and drag to the right. To maximize the text interface, drag to the left (PC version only). Rotate the 3D model by using the blue arrow buttons centered under it. The inner buttons rotate the 3D model step by step and the outer buttons rotate it continuously. Strip away anatomical layers, from deep to superficial, using the layer slider centered under the rotation arrows. Change the view of the 3D model by choosing additional views from the drop-down menu located on the lower right-hand side of the model interface.

To help you get started, Help balloons are available. Move the mouse over any button and its function will be revealed. (For a Mac, enable this by selecting Help balloons from the Help menu.) In-depth help is available from the Help menu.